# Encyclopedia of Environmental Issues
# Preservation and Wilderness Issues

# Encyclopedia of Environmental Issues
## Preservation and Wilderness Issues

*Editor*
**Craig W. Allin**
*Cornell College*

SALEM PRESS
A Division of EBSCO Publishing, Ipswich, Massachusetts

Cover photo:
*Bull Bison on Prairie.* (© Tom Bean/Corbis)

Copyright © 2011, by Salem Press, A Division of EBSCO Publishing, Inc.
All rights in this book are reserved. No part of this work may be used or reproduced in any manner whatsoever or transmitted in any form or by any means, electronic or mechanical, including photocopy, recording, or any information storage and retrieval system, without written permission from the copyright owner except in the case of brief quotations embodied in critical articles and reviews or in the copying of images deemed to be freely licensed or in the public domain. For information address the publisher, Salem Press, at csr@salempress.com.

ISBN: 978-1-42983-672-2

# Table of Contents

Contributors . . . . . . . . . . . . . . . . . vii

Alaska National Interest Lands Conservation Act . . 1
Antarctic and Southern Ocean Coalition . . . . . 2
Antiquities Act . . . . . . . . . . . . . . . . . 3
Aquaculture . . . . . . . . . . . . . . . . . . 4
Arctic National Wildlife Refuge . . . . . . . . . 6
Aswan High Dam . . . . . . . . . . . . . . . . 8

Beaches . . . . . . . . . . . . . . . . . . . . 10
Biosphere reserves . . . . . . . . . . . . . . 11
Bonn Convention on the Conservation of
 Migratory Species of Wild Animals . . . . . . 12

Carter, Jimmy . . . . . . . . . . . . . . . . . 13
Conservation . . . . . . . . . . . . . . . . . 14
Conservation movement . . . . . . . . . . . . 16
Conservation Reserve Program . . . . . . . . . 19
Controlled burning . . . . . . . . . . . . . . 20
Convention Relative to the Preservation of
 Fauna and Flora in Their Natural State . . . . 21
Cross-Florida Barge Canal . . . . . . . . . . . 22

Debt-for-nature swaps . . . . . . . . . . . . . 24
Deforestation . . . . . . . . . . . . . . . . . 25

Ecotourism . . . . . . . . . . . . . . . . . . 29
Endangered species and species protection
 policy . . . . . . . . . . . . . . . . . . . . 31
Environmental impact assessments and
 statements . . . . . . . . . . . . . . . . . 36
Environmental law, international . . . . . . . . 37
European Diploma of Protected Areas . . . . . . 41

Fish and Wildlife Service, U.S. . . . . . . . . . 42

Gila Wilderness Area . . . . . . . . . . . . . 43
Glen Canyon Dam . . . . . . . . . . . . . . . 44
Global ReLeaf . . . . . . . . . . . . . . . . . 46
Grand Canyon . . . . . . . . . . . . . . . . 47
Grand Coulee Dam . . . . . . . . . . . . . . 50
Great Swamp National Wildlife Refuge . . . . . 51
Green Plan . . . . . . . . . . . . . . . . . . 52
Greenbelts . . . . . . . . . . . . . . . . . . 53

Habitat destruction . . . . . . . . . . . . . . 54
Hetch Hetchy Dam . . . . . . . . . . . . . . 56

Indigenous peoples and nature
 preservation . . . . . . . . . . . . . . . . 58
International Union for Conservation
 of Nature . . . . . . . . . . . . . . . . . . 61

John Muir Trail . . . . . . . . . . . . . . . . 62

Kings Canyon and Sequoia national parks . . . . 63

Land-use planning . . . . . . . . . . . . . . 65
Leopold, Aldo . . . . . . . . . . . . . . . . . 67
Logging and clear-cutting . . . . . . . . . . . 68
Lovejoy, Thomas E. . . . . . . . . . . . . . . 70

Mather, Stephen T. . . . . . . . . . . . . . . 71
Migratory Bird Act . . . . . . . . . . . . . . 72
Mine reclamation . . . . . . . . . . . . . . . 73

Naess, Arne . . . . . . . . . . . . . . . . . . 75
National forests . . . . . . . . . . . . . . . . 76
National Park Service, U.S. . . . . . . . . . . . 78
National parks . . . . . . . . . . . . . . . . 80
National Trails System Act . . . . . . . . . . . 83
Nature Conservancy . . . . . . . . . . . . . . 85
Nature preservation policy . . . . . . . . . . . 86
Nature reserves . . . . . . . . . . . . . . . . 90

Preservation . . . . . . . . . . . . . . . . . 92

Ramsar Convention on Wetlands of
 International Importance . . . . . . . . . . 93
Recycling . . . . . . . . . . . . . . . . . . . 95
Roadless Area Conservation Rule . . . . . . . . 98
Roosevelt, Theodore . . . . . . . . . . . . . . 99
Rowell, Galen . . . . . . . . . . . . . . . . . 100

*Scenic Hudson Preservation Conference v.*
 *Federal Power Commission* . . . . . . . . . 101
Serengeti National Park . . . . . . . . . . . . 103
Sierra Club . . . . . . . . . . . . . . . . . . 105
*Sierra Club v. Morton* . . . . . . . . . . . . . 107
Soil conservation . . . . . . . . . . . . . . . 109

| | |
|---|---|
| Sustainable agriculture . . . . . . . . . . . . . . 110 | Wilderness Society . . . . . . . . . . . . . . . . 128 |
| Sustainable development . . . . . . . . . . . . 114 | Wildfires . . . . . . . . . . . . . . . . . . . . . . . . . 129 |
| Sustainable forestry . . . . . . . . . . . . . . . . 116 | Wildlife refuges . . . . . . . . . . . . . . . . . . 131 |
| | World Heritage Convention . . . . . . . . . . 134 |
| Tellico Dam . . . . . . . . . . . . . . . . . . . . . . 117 | World Summit on Sustainable Development . . 136 |
| | |
| United Nations Convention to Combat Desertification . . . . . . . . . . . . . . . . . . 118 | Yellowstone National Park . . . . . . . . . . . 138 |
| | Yosemite Valley . . . . . . . . . . . . . . . . . . 139 |
| Watt, James . . . . . . . . . . . . . . . . . . . . . 119 | |
| Wetlands . . . . . . . . . . . . . . . . . . . . . . . . 120 | *Zapovednik* system . . . . . . . . . . . . . . . . 141 |
| Wild and Scenic Rivers Act . . . . . . . . . . 123 | |
| Wilderness Act . . . . . . . . . . . . . . . . . . . . 125 | Bibliography . . . . . . . . . . . . . . . . . . . . . 143 |
| Wilderness areas . . . . . . . . . . . . . . . . . . 126 | Category Index . . . . . . . . . . . . . . . . . . 145 |
| | Index . . . . . . . . . . . . . . . . . . . . . . . . . . . 147 |

# Contributors

Emily Alward
*College of Southern Nevada*

Ruth Bamberger
*Drury College*

David Landis Barnhill
*Guilford College*

David Barratt
*Montreat College*

Alvin K. Benson
*Utah Valley University*

Cynthia A. Bily
*Macomb Community College*

Margaret F. Boorstein
*C. W. Post College of Long Island University*

Lakhdar Boukerrou
*Florida Atlantic University*

Victoria M. Breting-García
*Houston, Texas*

Robert E. Carver
*University of Georgia*

Frederick B. Chary
*Indiana University Northwest*

Thomas Clarkin
*University of Texas at San Antonio*

Kathryn A. Cochran
*Longview Community College*

Roy Darville
*East Texas Baptist University*

Joseph Dewey
*University of Pittsburgh*

Gordon Neal Diem
*ADVANCE Education and Development Institute*

John M. Dunn
*Ocala, Florida*

Daniel G. Graetzer
*University of Washington Medical Center*

Jerry E. Green
*Miami University*

William Crawford Green
*Morehead State University*

Wendy C. Hamblet
*North Carolina A&T State University*

Michael S. Hamilton
*University of Southern Maine*

Thomas E. Hemmerly
*Middle Tennessee State University*

Howard V. Hendrix
*California State University, Fresno*

Jane F. Hill
*Bethesda, Maryland*

Joseph W. Hinton
*Portland, Oregon*

Louise D. Hose
*Westminster College*

Allyson Leigh Hughes
*Michigan State University*

Allan Jenkins
*University of Nebraska at Kearney*

Karen N. Kähler
*Pasadena, California*

Jamie Michael Kass
*University of California, Berkeley*

Samuel V. A. Kisseadoo
*Hampton, Virginia*

Eugene Larson
*Los Angeles Pierce College*

Donald W. Lovejoy
*Palm Beach Atlantic University*

Steven B. McBride
*West Virginia University*

Nancy Farm Männikkö
*Centers for Disease Control and Prevention*

Laurence W. Mazzeno
*Alvernia College*

M. Marian Mustoe
*Eastern Oregon University*

Martin A. Nie
*University of Pittsburgh at Bradford*

P. S. Ramsey
*Brighton, Michigan*

C. Mervyn Rasmussen
*Renton, Washington*

Charles W. Rogers
*Southwestern Oklahoma State University*

Carol A. Rolf
*Rivier College*

Robert M. Sanford
*University of Southern Maine*

Elizabeth D. Schafer
*Loachapoka, Alabama*

Rose Secrest
*Chattanooga, Tennessee*

Adam B. Smith
*University of California, Berkeley*

Dion Stewart
*Adams State College*

Toby Stewart
*Duluth, Georgia*

Hubert B. Stroud
*Arkansas State University*

William R. Teska
*Furman University*

Donald J. Thompson
*California University of Pennsylvania*

Megan E. Watson
*Duke University*

Thomas A. Wikle
*Oklahoma State University*

Scott Wright
*University of St. Thomas*

Lisa A. Wroble
*Redford Township District Library*

Michele Zebich-Knos
*Kennesaw State University*

# Alaska National Interest Lands Conservation Act

CATEGORIES: Treaties, laws, and court cases; land and land use; preservation and wilderness issues

THE LAW: U.S. law intended to resolve conflicts regarding landownership in Alaska among Native Alaskans and the federal and state governments

DATE: Enacted on December 2, 1980

SIGNIFICANCE: The Alaska National Interest Lands Conservation Act more than tripled the amount of land protected as wilderness in the United States, but it did not lead to the resolution of all conflicts over Alaska land use.

The 1959 Alaska Statehood Act allowed the state government to choose 41.5 million hectares (102.5 million acres) of federal land from that part of the public domain not reserved for parks or other designated use. The land-selection process proceeded slowly, complicated by such issues as Native Alaskan claims, which were not clarified in the statehood bill. The discovery of oil fields in northern Alaska prompted passage of the 1971 Alaska Native Claims Settlement Act (ANCSA), which extinguished Native Alaskan land claims and allowed for construction of the Trans-Alaska Pipeline. To secure support for the ANCSA from conservation groups, the U.S. Congress included a provision in the bill that allowed the secretary of the interior to withdraw up to 32.4 million hectares (80 million acres) from the public domain for the establishment of conservation units, including national parks and wilderness areas. Congress would make the final determination regarding the final disposition of withdrawn lands. In order to provide a forum for cooperative planning on the issue, Congress created the Joint Federal-State Land Use Planning Commission.

An intense political battle ensued. Federal agencies, including the Forest Service and the National Park Service, competed for the authority to manage the withdrawn lands. Environmentalists favored agencies that would limit development, while Alaska state officials threw their support behind agencies that might prove willing to allow multiple use and the exploitation of natural resources for economic gain. After years of contentious debate, in 1979 the House of Representatives passed a bill establishing 51.4 million hectares (127 million acres) of conservation units, 26.3 million hectares (65 million acres) of which were designated wilderness areas. The Senate version of the bill significantly reduced the amount of land, and the legislation did not become law. The following year, the election of Ronald Reagan to the U.S. presidency and a Republican majority to the Senate convinced House members that they had no alternative but to accept the Senate version of the bill, which passed in 1980 and became known as the Alaska National Interest Lands Conservation Act (ANILCA).

ANILCA set aside 42.1 million hectares (104 million acres) of land as conservation units, with 23 million hectares (57 million acres) of wilderness. The National Park Service received 17.8 million hectares (44 million acres), the Fish and Wildlife Service 20.2 million hectares (50 million acres), and the Forest Service 1 million hectares (2.5 million acres). The Bureau of Land Management, the federal agency most amenable to development, received a patchwork of marginal lands that no other agency desired.

The act more than tripled the amount of land protected as wilderness in the United States, and its provisions placed 75 percent of national park land in Alaska. It did not solve the conflicts over Alaska land use, however. The state contested some land withdrawals. The act allowed for the construction of pipelines and roadways through conservation units. Moreover, ANILCA contained no measures that protected the habitats of migratory animals. Finally, the bill included provisions that allowed for the future exploitation of mineral resources, including oil, should both the Congress and the president deem it necessary.

*Thomas Clarkin*

FURTHER READING

Haycox, Stephen W. *Frigid Embrace: Politics, Economics, and Environment in Alaska.* Corvallis: Oregon State University Press, 2002.

Nelson, Daniel. *Northern Landscapes: The Struggle for Wilderness Alaska.* Washington, D.C.: Resources for the Future, 2004.

Ross, Ken. *Pioneering Conservation in Alaska.* Boulder: University Press of Colorado, 2006.

# Antarctic and Southern Ocean Coalition

CATEGORIES: Organizations and agencies; preservation and wilderness issues; ecology and ecosystems

IDENTIFICATION: Global coalition of nongovernmental organizations with a common commitment to preserve the lands of the Antarctic region and the southern oceans in perpetuity as a wilderness area

DATE: Founded in 1978

SIGNIFICANCE: The Antarctic and Southern Ocean Coalition is the only nongovernmental organization that serves to advocate for environmental protection and reform for the Antarctic region. It plays an important role in support of the region by monitoring the Atlantic Treaty System.

The Antarctic and Southern Ocean Coalition (ASOC) was established by the Friends of the Earth International, the World Wildlife Fund office of New Zealand, and other environmental organizations to monitor the Atlantic Treaty System, to implement its environmental protocols, and to provide expert witness in Antarctic affairs. ASOC seeks to advance the preservation and protection of the fragile ecosystems of the Antarctic continent and its surrounding waters. It reports regularly to key international organizations serving mandates to protect the Antarctic environment and its ecosystems from illegal hunting and fishing practices and to advocate for sustainable resource management, including waste disposal and habitat protection.

World Wildlife Fund founder Sir Peter Scott visited Antarctica in 1966 in the wake of successive international expeditions to the South Pole in the twentieth century, notably during the International Geophysical Year of 1957-1958. Prior to that time the southern Antarctic region and its complex relationship to global ecosystems were poorly understood. Despite regularly coordinated International Polar Year efforts, data systems were not sufficient to bring into clear focus the singular contributions of the southern oceans and icy landmasses of the Antarctic continent to the health of planetary biodiversity and the cosmography of global temperatures and climate. The critical roles of the Antarctic Convergence and the Antarctic Circumpolar Current in maintaining powerful hydraulic and chemical processes within the Pacific, Atlantic, and Indian oceans are of particular interest to international climatologists concerned with the possible consequences of global warming on sea levels and acidification.

Several international protocols are in place to preserve the Antarctic region in perpetuity as a pristine wilderness and global commons for international research. As an outcome of the International Geophysical Year of 1957-1958, twelve nations established research stations in the Antarctic, and on December 1, 1959, they signed the Antarctic Treaty proposed by the United States. By 2010, forty-seven international participants had agreed to honor the treaty's commitment to collaborative research and environmental protection while proscribing military operations on the continent. Twenty-eight nations serve as consultative parties with authority to make decisions on behalf of the treaty.

ASOC works closely with the Antarctic Treaty Secretariat and attends meetings sponsored by other international organizations, including the International Maritime Organization and the International

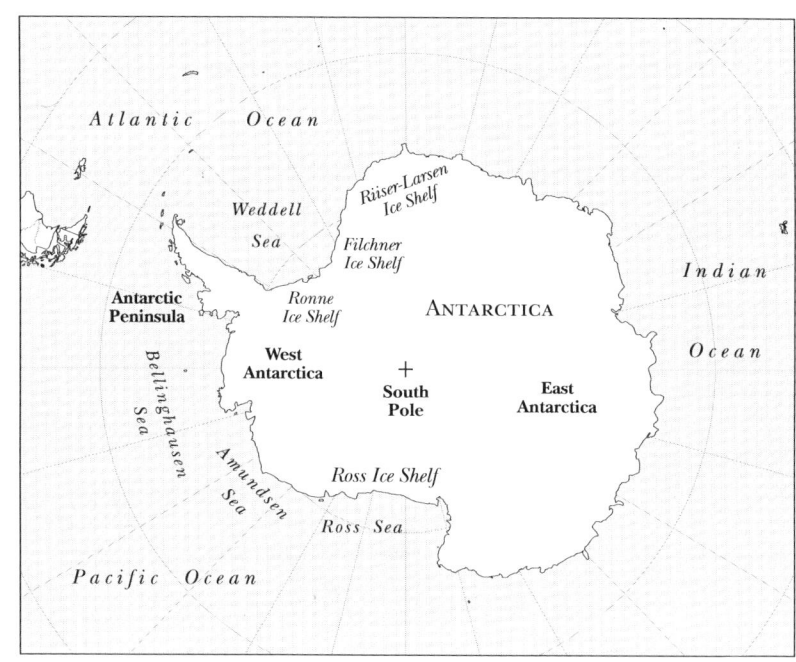

Whaling Commission, as well as the bodies overseeing the Convention on the Conservation of Antarctic Marine Living Resources and the Agreement on the Conservation of Albatrosses and Petrels. Major ASOC campaigns have included the Antarctic Krill Conservation Project, the Southern Ocean Whale Sanctuary, and protection for the Ross Sea.

As an outcome of the 2007-2008 International Polar Year, the Scientific Committee on Antarctic Research (SCAR) developed a network of databases cataloging the extraordinary biodiversity of the South Pole region. The brainchild of Claude De Broyer and Bruno Danis at the Royal Belgian Institute of Natural Sciences, the Marine Biodiversity Information Network (SCAR-MarBIN) is a collaborative effort by hundreds of scientists worldwide to establish the Register of Antarctic Marine Species, the first complete online resource of its kind. This system collates data from more than one hundred international databases to create detailed mappings of the delicate ecology of Antarctica. These images provide valuable information for scientists seeking to understand and analyze the biological equilibrium of the region and the implications of global climate change on life-forms unique to its lands and waters.

*Victoria M. Breting-García*

FURTHER READING

Bargagli, R. *Antarctic Ecosystems: Environmental Contamination, Climate Change, and Human Impact*. New York: Springer, 2005.

Joyner, Christopher C. *Governing the Frozen Commons: The Antarctic Regime and Environmental Protection*. Columbia: University of South Carolina Press, 1998.

Steig, Eric J., et al. "Warming of the Antarctic Ice-Sheet Surface Since the 1957 International Geophysical Year." *Nature* 457 (January 22, 2009): 459-462.

Triggs, Gillian D., ed. *The Antarctic Treaty Regime: Law, Environment, and Resources*. 1987. Reprint. New York: Cambridge University Press, 2009.

Turner, John, et al., eds. *Antarctic Climate Change and the Environment*. Cambridge, England: Scientific Committee on Antarctic Research, 2009.

## Antiquities Act

CATEGORIES: Treaties, laws, and court cases; preservation and wilderness issues

THE LAW: U.S. federal law empowering the president to set aside as national monuments public lands deemed to be of historic or scientific importance

DATE: Enacted on June 8, 1906

SIGNIFICANCE: The Antiquities Act was one of the earliest pieces of U.S. federal legislation aimed at historic and cultural preservation. Since passage of the act, U.S. presidents have used it more than one hundred times to protect natural phenomena as well as human-built sites of cultural interest.

The original impetus for the Antiquities Act of 1906 was public reaction to the widespread looting of prehistoric Native American ruins in the Southwest during the late nineteenth century. One of the individuals seeking some form of federal protection for these sites was anthropologist Edgar Lee Hewett, and Congressman John F. Lacey, a Republican from Iowa, joined in the effort during the early part of the century. President Theodore Roosevelt signed the resulting legislation into law in June of 1906, and three months later he used the act for the first time to establish the Devils Tower National Monument in Wyoming.

All subsequent U.S. presidents have exercised the powers given them by the law to establish a wide variety of protected sites, from natural phenomena such as the Grand Canyon in Arizona and the Muir Woods in California to historic sites such as the Statue of Liberty and the birthplace of George Washington. The environmental significance of the act was expanded during the presidency of Jimmy Carter when it was used (in 1978) to establish a series of fifteen national monuments preserving a total of 22.7 million hectares (56 million acres) of wilderness land in Alaska, protecting the land from oil drilling and other forms of commercial development.

*Scott Wright*

# Aquaculture

CATEGORY: Agriculture and food

DEFINITION: Production of marine and freshwater food sources in controlled, farmlike environments in ponds, canals, lakes, and confined coastal areas

SIGNIFICANCE: Although aquaculture provides a good source of protein for the demands of a growing world population, as well as an economic endeavor for some, like many forms of farming it is not without its negative environmental impacts. Residues from fish wastes, pollution, coastal erosion, and impacts on adjacent species are some of the challenges to sustainable forms of aquaculture.

Aquaculture is one of the fastest-growing forms of food production in the world. It includes the processes of propagating, raising, and processing marine and freshwater food sources such as fish, shellfish, and even kelp. In some cases by-products of the production of these commodities produce fertilizers and feeds for other animals. Ornamental products such as cultured pearls are also produced within aquacultural systems.

The practice of aquaculture is not new. Fish farming has been a part of the cultures of Pacific islanders and Southeast Asian peoples for centuries. Early forms of aquaculture were quite rudimentary compared with modern techniques, however; they provided foods and other products on a limited scale, mostly for local villages. In the twenty-first century, sophisticated aquaculture systems control the complete life cycles of the fish and other animals under cultivation. In many of these modern facilities, automated feeding and harvesting systems alleviate the need for labor-intensive practices, reducing the costs of production. In some cases, however, because of the nature of the commodities, intensive human labor is still required. Oysters, for example, are harvested from beds and then must be opened by hand, a process known as shucking, so that the integrity of the shells is maintained for the seeding of a future crop.

One of the elements that has spurred the development of aquaculture is the decline in production in fisheries in the open oceans owing to overfishing. According to the United Nations Food and Agriculture Organization (FAO), all major fishing regions in the world are now being fished intensely. The needs of the growing world population for protein and water pollution affecting natural sources of seafoods have also contributed to the growth of aquaculture.

INDUSTRY GROWTH

Commercial aquaculture ventures have been established in nations throughout the world. The majority of these operations utilize naturally existing water sources in coastal regions, canals, or rivers; others are built at sites where traditional agriculture could not take place. An ideal aquaculture site has certain conditions that are favorable to the healthful management of the stocks of fish and other animals produced. These include clean water, appropriate water temperatures and levels of dissolved gases, good site topography, and food sources.

In the United States aquaculture operations are found in coastal regions and in dry interior areas, where ponds or rivers are used. Clams and oysters are harvested from beds in the marine environment of Washington State's Puget Sound, for example, while in the eastern interior of the state, the Columbia River provides a freshwater source for the raising of salmon, trout, and steelhead, all for commercial uses. In dry west Texas, gravel pits filled with saline waters were some of the first ponds used to cultivate shrimp, an industry that now supplies millions of pounds annually from ponds filled with natural saline water. The state of Virginia has developed a large aquaculture industry, growing clams and oysters in hatcheries.

Crawfish, tilapia, trout, shrimp, cod, clams, oysters, frogs, and catfish are just some of the products found in aquacultural operations in the United States. According to the U.S. National Oceanic and Atmospheric Administration, shellfish account for two-thirds of the total output of the U.S. aquaculture industry; salmon accounts for 25 percent, and shrimp for about 10 percent. Catfish is one of the most popular farmed fish in the United States.

The monetary value of worldwide aquaculture production is some $70 billion per year. FAO has estimated that almost half of the fish eaten in the world, nearly 45 billon tons of fish, is produced in aquacultural environments rather than caught in the open seas. According to FAO, the top ten aquaculture-producing nations are, in descending order, China, India, Vietnam, Thailand, Indonesia, Bangladesh, Japan, Chile, Norway, and the United States. Even with its tenth-place ranking, the U.S. aquaculture industry contributes about $1 billion annually to the nation's domestic economy.

## Environmental Concerns

Aquaculture operations have a number of negative environmental impacts. Among these is that they generally require the installation of a considerable amount of technology, and thus disturb existing habitats, and they use large amounts of energy. Feed must be supplied to penned fish, and waste from the fish must somehow be managed. Diseases must be controlled, not only within the fish and other seafood stocks but also with respect to the possible spread of disease to adjacent native fish and other organisms outside the aquaculture pens or ponds. A particularly controversial element of aquaculture involves the genetic modification of many forms of fish stocks to enhance their ability to grow quickly under confined conditions. Critics of these practices question the safety of eating such fish and have also voiced concerns that if genetically modified species escape their aquaculture pens they may introduce unforeseeable problems into natural fish populations.

In many aquacultural settings, self-contained ecosystems are set up to supply nutrients for fish stocks. In some cases sewage and even commercial fertilizers have been used in such ecosystems to stimulate the production of phytoplankton, which in turn is eaten by zooplankton and bottom dwellers. These are then eaten by fish stocks within the pen or pond. Unless a consistent flow of water can be maintained through the pen area where the fish stock is raised, the bottom can become silted up with waste. In confined estuaries this condition can have negative effects on fish outside the aquacultural operation. It has been suggested that the decay of fecal matter at the bottom of aquaculture pens and in areas where concentrated fish-rearing operations are conducted can contribute to the depletion of oxygen levels in the water near the bottom. This is especially the case if water currents are not efficient enough to purge the buildup of waste at the bottoms of these ponds and pens.

Shrimp, a major commodity grown aquaculturally, is a good example of some other environmental concerns. Aquaculture operations sometimes use antibiotics to control the diseases that shrimp are prone to acquire when they are being raised in confined areas. During the early part of 2000, the European Union banned shipments of shrimp containing chloramphenicol, an antibiotic used on shrimp that has been found to be dangerous for human health. In addition, fishing communities and coastal residents have been critical of shrimp aquaculture, asserting that it reduces fish catches because it has negative impacts on coastal habitats and because it uses wild fish as food for the shrimp being raised.

Additional concerns about the negative environmental impacts of aquaculture have to do with the use of pesticides and other chemicals in the management of unwanted organisms in aquacultural waters. To control infestations of fish lice, for example, aquaculture operations often disperse pesticides into the water. Other chemicals are often used to cut down on algae growth on the nets and floats used in pen operations.

Within food webs or trophic levels, energy flows from lower to higher levels. If a larger fish eats a smaller fish, not all of the energy available from the smaller fish is passed on to the larger. This transference of energy is known as ecological efficiency. Aquaculture can exhibit very low levels of ecological efficiency. For example, farm-raised salmon are reared on food made from other fish, which are caught and ground up before being fed to the salmon as pellets or in meal form. It has been estimated that it takes 10 grams (0.35 ounce) of feed to produce 1 gram (0.035 ounce) of salmon. Considering the pressure already existing on wild fish stocks, it has

### U.S. Aquaculture Production, 2006

| | Thousands of Pounds | Metric Tons | Thousands of Dollars |
|---|---|---|---|
| *Finfish* | | | |
| Baitfish | — | — | 38,018 |
| Catfish | 566,131 | 256,795 | 498,820 |
| Salmon | 20,726 | 9,401 | 37,439 |
| Striped bass | 11,925 | 5,409 | 30,063 |
| Tilapia | 18,738 | 8,500 | 32,263 |
| Trout | 61,534 | 27,912 | 67,745 |
| *Shellfish* | | | |
| Clams | 12,564 | 5,699 | 72,783 |
| Crawfish | 80,000 | 36,288 | 96,000 |
| Mussels | 962 | 436 | 4,990 |
| Oysters | 13,711 | 6,219 | 92,602 |
| Shrimp | 8,037 | 3,646 | 18,684 |
| *Miscellaneous* | — | — | 254,738 |
| **Totals** | **794,328** | **360,305** | **1,244,145** |

Source: Data from the National Oceanic and Atmospheric Administration, National Marine Fisheries Association.
Note: Miscellaneous includes ornamental and tropical fish, alligators, algae, aquatic plants, eels, scallops, crabs, and others.

been argued that this low level of ecological efficiency is a considerable detriment for the economy of aquaculture. In contrast, some kinds of aquaculture operations, such as catfish farms, are relatively ecologically efficient, and catfish, which are herbivorous, require no fish-based foods.

Aquaculture operations are also sometimes negatively affected by environmental conditions that come from outside their confines. Along coastlines and in low-lying areas, for example, pollution from agricultural irrigation runoff can cause fish kills such as those that have taken place in aquaculture installations in Indonesia, Malaysia, and the Philippines. Washington State enacted a law to control the flow of manure effluent from dairy farms into the Nooksack River. Prior to the control of this pollution, coastal clam beds managed by the Lummi Nation were decimated; this aquacultural operation recovered after passage of the law.

*M. Marian Mustoe*

FURTHER READING

Jahncke, Michael L., et al., eds. *Public, Animal, and Environmental Aquaculture Health Issues.* Hoboken, N.J.: John Wiley & Sons, 2002.

Mathias, Jack A., Anthony T. Charles, and Hu Baotong, eds. *Integrated Fish Farming.* Boca Raton, Fla.: CRC Press 1997.

Miller, G. Tyler, Jr., and Scott Spoolman. "Food, Soil, and Pest Management." In *Living in the Environment: Principles, Connections, and Solutions.* 16th ed. Belmont, Calif.: Brooks/Cole, 2009.

Pilay, T. V. R., and M. N. Kutty. *Aquaculture: Principles and Practices.* 2d ed. Ames, Iowa: Blackwell, 2005.

Stickney, Robert R. *Aquaculture: An Introductory Text.* 2d ed. Cambridge, Mass.: CABI, 2009.

## Arctic National Wildlife Refuge

CATEGORIES: Places; animals and endangered species

IDENTIFICATION: Large area of land in northeastern Alaska set aside by the U.S. government for the preservation and protection of wildlife

DATES: Established in 1960; expanded in 1980

SIGNIFICANCE: The Arctic National Wildlife Refuge provides a home for a great many species of marine mammals, birds, and terrestrial animals. Its isolation, biodiversity, and protected status make it an important sanctuary for threatened species such as the polar bear. The possibility that oil exploration could be undertaken in a section of the refuge is a topic of ongoing debate.

Even before Alaska gained statehood in 1959, a movement was under way in the National Park Service and among conservationists to protect a small part of northeastern Alaska permanently from development and commercial interests. The Arctic National Wildlife Range was established as a federally protected area in 1960 by the U.S. secretary of the interior, Fred A. Seaton, in order to preserve the uniqueness of the wilderness and its wildlife. In 1980 passage of the Alaska National Interest Lands Conservation Act (ANILCA) expanded the protected area from less than 3.6 million hectares (9 million acres) to approximately 7.3 million hectares (18 million acres) and renamed it the Alaska National Wildlife Refuge (ANWR).

The remoteness and protected status of the refuge have limited the human impact on the environment there. It provides important habitat for marine life, which includes seals and whales as well as fish and seabirds, and terrestrial animals such as caribou, wolves, and the three North American species of bear. The refuge is home to many animals that flourish during the short Arctic summer, such as a variety of insects, and a safe haven where migratory birds can rest while moving south each year. The refuge has been a subject of controversy since the 1970's, and supporters of the area's protected status have fought off many attempts to open it up to oil and gas development.

GEOGRAPHY AND WILDLIFE

The Arctic National Wildlife Refuge, sometimes called the Arctic Refuge, has greater species diversity than any other protected area in the Arctic Circle. The northern boundary of the refuge is the coastline of the Beaufort Sea and has habitats typical of coastal regions, such as river deltas, barrier islands, and salt marshes. These support wildlife that includes migratory seabirds and varieties of fish; in the summer caribou herds travel there to give birth and raise their young. Also during the summer, in addition to the many insects that thrive there, tens of thousands of migratory birds feed and rest on the coastal plain before traveling south. During the winter months, polar bears create birthing dens and hunt seals on the sea ice that grows along the coast.

Further inland, the coastal habitats give way to a plain, dotted with small lakes and braided rivers, that gradually moves upland to the Brooks Range. The Brooks Range region is made up of the foothills, valleys, and mountains north and south of the mountains and provides habitat for, among others, wolves, ducks, and birds of prey such as falcons and eagles. The region of the refuge on the south side of the Brooks Range consists mainly of the boreal forest characteristic of inland Alaska. Year-round residents of this forest include grizzly bears, lynx, wolverines, and moose; caribou herds spend the winter there, and migratory birds breed there during the spring and summer. On this side of the Brooks Range, the rivers that flow south to the Yukon River, the wetlands of the region, and the forest canopy provide a wide variety of habitats and food for many different species, from fish to mammals to birds.

### PROTECTION AND CONTROVERSY

In 1968 the discovery of oil in Alaska's Prudhoe Bay, which is only sixty miles west of the western border of ANWR, began a political controversy that has not abated. When the Alaska National Interest Lands Conservation Act enlarged the refuge in 1980, it also left open the possibility of oil and gas exploration and exploitation on 607,000 hectares (1.5 million acres) of the refuge's coastline and coastal plain, which are areas of great biodiversity and are of vital importance to many species. Section 1002 of ANILCA allowed an opening for oil exploration in ANWR, but only if mandated by Congress and the president; this provision of the law was an unhappy compromise that each side vowed to change in its own favor as soon as possible.

Since 1980 many attempts have been made to open what became known as the 1002 Area to exploration and drilling, particularly under Republican presidential administrations and by Republican members of Congress. Alaskan local, state, and national politicians have been among the greatest proponents of drilling in the refuge and have generally been supported by their constituents. Taxes and royalties from oil drilling in Prudhoe Bay provide a large amount of revenue for Alaska annually, and each state resident gets a share in the form of a yearly check; in 2000 the amount of this payment was reportedly about $1,900 per person. The refuge is federal land, however, so the issue must be decided in Washington, D.C., not in Alaska.

During President George W. Bush's two terms in office, moves to open the refuge to oil drilling occurred repeatedly in the Republican-controlled Congress, at first in specific proposals to allow drilling and then, when that approach failed, through attachments to major bills, such as defense-spending and energy bills, or additions to federal budgets. Whether enough oil actually exists in the 1002 Area to make drilling economically worthwhile to the United States and acceptable to many Americans remains unknown. What is clear is that the issue of drilling in the Arctic National Wildlife Refuge exemplifies the global dilemma between the need for energy and the need for a healthy planet.

*Megan E. Watson*

### Further Reading

Kaye, Roger. *Last Great Wilderness: The Campaign to Establish the Arctic National Wildlife Refuge.* Fairbanks: University of Alaska Press, 2006.

Standlea, David M. *Oil, Globalization, and the War for the Arctic Refuge.* Albany: State University of New York Press, 2006.

Vaughn, Jacqueline. *Conflicts over Natural Resources: A Reference Handbook.* Santa Barbara, Calif.: ABC-CLIO, 2007.

## Aswan High Dam

CATEGORY: Preservation and wilderness issues
IDENTIFICATION: Dam on the Nile River in Egypt
DATE: Completed in 1970
SIGNIFICANCE: The Aswan High Dam supplies electricity and has helped control seasonal flooding, but it has had several negative environmental effects, including loss of fertility on downriver floodplains, increased erosion, and increased incidence of earthquakes.

The Aswan High Dam was built with the aid of Soviet engineers on the Nile River approximately 960 kilometers (600 miles) south of Cairo, Egypt, between 1960 and 1970 at a cost of US$1 billion. The dam lies 7 kilometers (4.5 miles) south of the city of Aswan and several kilometers from a smaller dam constructed by British engineers between 1898 and 1902. The building of the High Dam followed the signing of the Nile Water Agreement between Egypt and Sudan in November of 1959.

The High Dam is a rock-fill structure with a core of impermeable clay. It measures 3,829 meters (12,562 feet) long, 111 meters (364 feet) high, 980 meters (3,215 feet) wide at the base, and 40 meters (131 feet) wide at the crest. The volume of material contained in the structure, 1.6 million cubic meters (56.5 million cubic feet), would be enough to construct seventeen Great Pyramids. The flow of the river's waters through the dam is via six tunnels, each controlled by a 230-ton gate.

The reservoir impounded by the High Dam is Lake Nasser, named for Egyptian president Gamal Abdel Nasser, who died the year the dam was completed. The portion of the reservoir that lies within Sudan, about 30 percent, is referred to as Lake Nubia. In total, the reservoir measures 499 kilometers (310 miles) in length and has a surface area of approximately 5,996 square kilometers (2,315 square miles) and 9,053 kilometers (5,625 miles) of shoreline. It averages 9.7 kilometers (6 miles) wide, with a maximum width of 16 kilometers (10 miles). Mean water depth is 70 meters (230 feet), maximum depth is 110 meters (360 feet), and the annual vertical fluctuation is 25 meters (82 feet). The reservoir contains enough water to irrigate more than 2.8 million hectares (7 million acres).

### ENVIRONMENTAL IMPACTS

The filling of Lake Nasser came with a high cost. Thousands of people were displaced, and natural habitats were significantly altered. A number of ancient temples and monuments, which abound in the region, were to be submerged beneath the rising waters of the reservoir, but some of these were saved—they were cut into large blocks and reassembled at higher locations. High evaporation rates and water loss to infiltration into the underlying permeable Nubian sandstone caused the filling of the reservoir to take much longer than anticipated. A significant amount of water-storage capacity has also been lost as sediment that has been carried in occupies a part of the reservoir's volume.

The benefits derived from the High Dam are principally hydroelectric power generation and the regulation of water flow along the lower Nile for flood control. Twelve turboelectric generators capable of producing 10 billion kilowatt-hours provide 40 percent of Egypt's electrical power. The storage of water in Lake Nasser not only provides flood control but also allows for the irrigation of additional land and the ability to grow multiple crops over the course of a year. Since the filling of the reservoir, a fishing industry has also developed.

The yearly floods of the Nile are the result of late-summer rains that fall in the plateau region to the south in Ethiopia. At peak floods, river volume may increase by as much as sixteen times. More than 100 million tons of soil are carried with the water each year. While the impoundment of water in Lake Nasser has probably saved the lower Nile Valley from disastrous floods and alleviated the effects of regional droughts, the loss of yearly increments of silt—with its associated nutrients—on the floodplain has led to a decline in the floodplain's fertility. Without this natural fertilization, the Egyptians have had to rely on increasing use

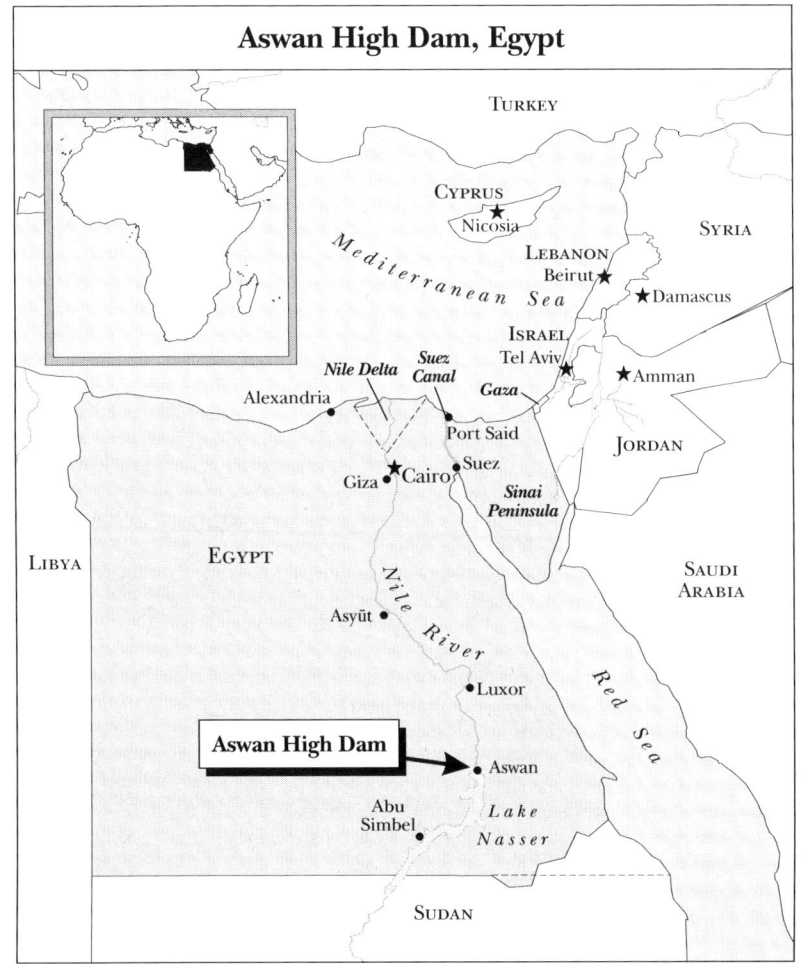

**Aswan High Dam, Egypt**

of artificial fertilizer. The floodwaters also provided a cleansing and draining action for the soil, preventing the accumulation of salts. Further, the floodwaters reduced the numbers of rats and disease-bearing snails. With the decline in flooding, incidences of disease have been on the increase.

Historically, the influx of sediment to the Nile Delta has replenished sediment lost to wave and current erosion at the Delta's margins. Since the construction of the Aswan High Dam, the front of the Delta is being eroded at a rate of 1.8 meters (6 feet) per year. As on the floodplain, the soil of the Delta, a region that has been farmed for more than seven thousand years, also shows evidence of declining fertility.

Since the High Dam traps 98 percent of the Nile sediment, water passing through the dam has an enhanced ability to erode. Consequently, downstream erosion has become a significant problem, scouring the riverbed and undermining riverbanks and bridge piers. In some instances, the increased erosion has also affected the delicate balance of water irrigation systems.

The effects of trapped sediment are not confined to the floodplain and the Nile Delta. Prior to construction of the dam, the river brought sediment and nutrients into the normally nutrient-poor eastern Mediterranean Sea. This provided for blooms of phytoplankton that formed the base of a food pyramid that included sardines and other commercial varieties of fish. When construction of the dam began, the sardine fishing industry in the Mediterranean declined significantly. From the late 1980's onward, however, a resurgence occurred in sardine fishing, as the filling of Lake Nasser allowed for increased river discharge and nutrient enhancement.

One other environmental consequence of the construction of the High Dam has been the occurrence of earthquakes in the region. These are related to stress that is placed on the earth's crust by the weight of the water impounded in Lake Nasser. A large shock of magnitude 5.6, for example, occurred on November 14, 1981. This was followed by aftershocks for a period of seven months.

*Donald J. Thompson*

**FURTHER READING**

Caputo, Robert. "Journey up the Nile." *National Geographic*, May, 1985.

Collins, Robert O. *The Nile*. New Haven, Conn.: Yale University Press, 2002.

Johnson, Irving, and Electra Johnson. "Yankee Cruises the Storied Nile." *National Geographic*, May, 1965.

Smith, Scot E. "The Aswan High Dam at Thirty: An Environmental Impact Assessment." In *Conservation, Ecology, and Management of African Fresh Waters*, edited by Thomas L. Crisman et al. Gainesville: University Press of Florida, 2003.

# Beaches

CATEGORY: Preservation and wilderness issues
DEFINITION: Sandy or rocky areas on the shores of bodies of water
SIGNIFICANCE: Beaches have significant value for recreational purposes, and important industries, such as those related to swimming and surfing, depend on them. Beaches worldwide are threatened by accelerating erosion, as well as by a projected rise in sea level.

Beaches are accumulations of loose sedimentary material found along the shorelines of bodies of water. They extend landward from the low-tide level to a place where there is a marked change in appearance or topography of the land surface, such as a cliff, a row of sand dunes, or human-made structures such as seawalls. The two major subdivisions of a beach are the backshore, which extends seaward from the foot of a cliff, dunes, or seawall; and the foreshore, which usually has a steeper slope that continues down to water level. The boundary between the backshore and the foreshore is generally marked by a change in slope or by a small scarp excavated by wave activity. The seaweed and other debris that accumulate at this point are known as the wrack.

The backshore is considered to be the inactive part of a beach because there is little evidence for wave activity here. It is the part of a beach most used by humans. Human-built additions such as volleyball nets, lifeguard stands, and campfire pits are found on the backshore; grasses or trailing vines may grow down from cliff, dunes, or seawall to extend partway across the backshore.

The foreshore is the active part of the beach. It is covered daily by the rise and fall of the tides; no grasses or trailing vines are found here because the waves and tides would sweep them away immediately. All traces of human activity, such as footprints and lost toys, are removed by the waves as well. Upon going out, the tide leaves the surface of this part of the beach perfectly smooth, broken only at its lower end by a slight dip known as the low-tide step. This is a gentle trough excavated by the breakers at low tide. Waves grind up shells and pebbles here in a so-called wave mill, and the resulting fragments become new materials for the beach. They grow finer-grained as the wind and waves carry them upward toward the dunes.

The widths of beaches vary greatly around the world. Beaches at the foot of cliffs or seawalls may be just a few feet wide, whereas those on low-relief coasts can extend seaward for many miles.

ACCELERATING BEACH EROSION

Oceanographers calculate that 70 percent of the world's beaches are being cut back by erosion, 10 percent are growing forward, and the rest remain unchanged. Erosion of a beach occurs when it loses more sediment than it gains from the sources that feed it. Losses may take place when beach materials are carried inland during storm surges, when they are carried seaward as big waves cut back the coast during severe storms, or when they are carried along the shoreline by the longshore current. This current is

---

### Currents That Affect Beaches

Longshore currents are those parallel to a seashore and extending from the shoreline through the breakers. They occur because waves approach coasts at a small angle and bend in the shallows. The speed of a longshore current is related to wave size and angle of approach to a shore. During quiet weather, longshore currents move at under half a mile per hour, but during storms they can move ten to twelve times that speed. The combined actions of waves and longshore currents transport a lot of sediment along shallows bordering a shore. Longshore currents move in either direction along a beach, depending on the direction of wave approach, itself a result of wind direction. Thus waves suspend sediment, and longshore currents transport it along beaches.

Tides, the regular rise and fall of sea level due to the gravitational fields of the moon and the sun, cause daily changes in ocean levels of 1 to 50 feet (0.30 to 15 meters). Tidal currents transport large amounts of sediment and erode rock, and tidal rise and fall distribute wave energy across shores by changing water depth. In estuaries, tides create the speeds needed to move sand. On open coasts—that is, on beaches—tides do not move the water fast enough for sediment transfer. However, the rise and fall of tides along open coasts indirectly affect sediment movement, because their landward movement or retreat causes shorelines to move. This changes the region where waves and longshore currents operate. Beach slope is also crucial, with gently sloped beaches having the largest shoreline changes during tide cycles.

present along coasts whenever waves strike the shoreline at an acute angle rather than coming in head-on.

During severe storms the loss of beach materials can be rapid; evidence for this is seen in collapsed seawalls and cliffed dunes on the backshore or in salt marsh deposits and tree stumps uncovered on the foreshore as the beach moves inland over a previously vegetated area. Aerial photographs are generally used to determine how much material has been removed from beaches, with recent photographs compared to photographs taken previously.

Beach erosion accelerated in the early twenty-first century because of several factors: the active removal of materials by waves, wind, and currents; rising sea level, which resulted in the migration of shorelines inland; and increasing human modification of the coasts. Humans' changes to coastlines that affect beach erosion include both the construction of seawalls, groins, offshore breakwaters, and jetties and the bringing in of new sand in a procedure known as beach nourishment. Seawalls are built at the backs of beaches to prevent erosion, but they may reflect the storm waves, causing them to carry sediment seaward. Groins, which are rock ribs extending seaward along the foreshore to trap sand on their up-drift sides, cause erosion on their down-drift sides, resulting in alternating scallops of erosion and deposition along the coast. Breakwaters, which are offshore structures built parallel to the coast, and jetties, which are walls extending seaward to stabilize inlets, both trap migrating sand, resulting in erosion on their down-drift sides. Even beach nourishment, a highly popular practice, has its negative effects. Because the sand that is added to beaches is dredged from offshore or trucked in from the land, it rarely has the same texture and composition of the original beach sand. The result is that this new sand usually erodes faster than the original sand that it replaces.

*Donald W. Lovejoy*

FURTHER READING

Bird, Eric. *Coastal Geomorphology: An Introduction.* 2d ed. Hoboken, N.J.: John Wiley & Sons, 2008.

Davis, Richard A., Jr., and Duncan M. FitzGerald. *Beaches and Coasts.* Malden, Mass.: Blackwell, 2004.

Neal, William J., Orrin H. Pilkey, and Joseph T. Kelley. *Atlantic Coastal Beaches: A Guide to the Ripples, Dunes, and Other Natural Features of the Seashore.* Missoula, Mont.: Mountain Press, 2007.

## Biosphere reserves

CATEGORY: Preservation and wilderness issues
DEFINITION: UNESCO-designated sites where the preservation of natural resources is integrated with research and the sustainable management of those resources
SIGNIFICANCE: The worldwide network of biosphere reserves is the only international network of protected areas that also emphasizes sustainable development and wise use of natural resources; thus these sites enable the examination and testing of the objective of integrating conservation and development.

Through discussions that started in 1970, the United Nations Educational, Scientific, and Cultural Organization (UNESCO) initiated the Man and the Biosphere Programme to establish sites where the preservation of natural resources would be integrated with research and the sustainable management of those resources. The first biosphere reserve was designated in 1976, and by May, 2009, 553 such sites existed across 107 countries. In 1995 UNESCO convened a conference in Spain and developed the Seville Strategy for Biosphere Reserves, which was designed to strengthen the international network and encourage the use of the sites for research, monitoring, education, and training. In the early years of the program, preservation was stressed. The adoption of the Seville Strategy by UNESCO emphasized the role of people in the use of their natural resources.

Each biosphere reserve contains a legally protected core area where there has been minimal disturbance by people. Only uses that are compatible with the preservation of biological diversity are permitted in the protected core. Surrounding the core is a managed-use or buffer zone; research and environmental education are examples of activities suitable for the buffer zone. Surrounding the buffer zone is a zone of cooperation or transition zone. The boundaries of the transition zone are loosely defined and often include local towns and communities. Economic activities such as farming, logging, mining, and recreation occur within the transition zone and are not restricted by the biosphere reserve.

Only the boundaries of the core area are legally defined. The designation of a biosphere reserve does not alter the legal ownership of the land or water that is included within its zones. UNESCO does not have jurisdiction over any nation's biosphere reserves. In many cases the areas within reserves reflect a mosaic of landownership, including federal, state, local, and private ownership. Even the core area may be privately owned, as long as it is managed for its preservation.

In the United States, the core areas of some biosphere reserves are within national parks, such as Glacier and Yellowstone, whereas other biosphere reserves are composed of clusters of core areas, such as the ten units within the California Coastal Range Biosphere Reserve. The management and administration of a biosphere reserve often involves a number of interested citizens, government agencies, and owners.

The worldwide network of biosphere reserves represents the only international network of protected areas that also emphasizes sustainable development and wise use of natural resources. Hence they are sites where the objective of integrating conservation and development can be examined, demonstrated, and tested. Research at these sites serves to solve practical problems in resource management.

*William R. Teska*

FURTHER READING

Hanna, Kevin S., Douglas A. Clark, and D. Scott Slocombe, eds. *Transforming Parks and Protected Areas: Policy and Governance in a Changing World.* New York: Routledge, 2008.

Sourd, Christine. *Explaining Biosphere Reserves.* Paris: UNESCO, 2004.

# Bonn Convention on the Conservation of Migratory Species of Wild Animals

CATEGORIES: Treaties, laws, and court cases; animals and endangered species

THE CONVENTION: International agreement protecting animal species that migrate across national borders

DATE: Opened for signature on June 23, 1979

SIGNIFICANCE: The Convention on the Conservation of Migratory Species of Wild Animals encourages the protection of migratory species by nations around the world. Most of the convention's success has been in fostering regional daughter agreements that protect particular migratory species of concern.

The Convention on the Conservation of Migratory Species of Wild Animals (also known as CMS or the Bonn Convention) arose as the result of recommendations made at the 1972 United Nations Conference on the Human Environment held in Stockholm, Sweden. Opened for signing in Bonn, Germany, in 1979, the treaty entered into force in 1983. By 2010 the convention had been signed by 113 nations, and many other nonsignatory nations were participating in CMS-sponsored agreements. The United Nations Environment Programme provides the CMS Secretariat.

The convention recognizes that the conservation status of a migratory species is vulnerable to threats in any states that species occupies or passes through. Appendix I to the convention lists endangered species for which concerted conservation action is required to ensure their survival. Parties to the convention (that is, signatory nations) are encouraged to forbid the taking of these species, to take steps to enhance their welfare, to remove obstacles to their migration, and to manage uses of the species by indigenous cultures. Appendix II lists migratory species that are not necessarily endangered but would benefit from conservation action.

CMS encourages states whose territories include the biogeographic ranges of migratory species to enter into one of four types of daughter agreements. In order of formality from least to most, these are designated as Action Plans, Memoranda of Understanding (MOUs), agreements (all lowercase in the CMS text), and AGREEMENTS (all capitals in the convention text).

Action Plans require the least of signatories and generally affirm that species of concern would benefit from research and conservation action and that parties will endeavor to engage in such action. Action Plans are also negotiated as components of the other kinds of agreements. Of the few stand-alone Action Plans that have been developed, one protects birds that use the migration routes within the Central Asian Flyway and another protects several species of African antelope.

MOUs are typically nonenforceable and do not require obligatory actions by parties, but they can help draw official and public attention to species of concern and result in conservation efforts on the species' behalf. The MOUs that have been negotiated under the auspices of CMS cover a range of species, from West African elephants to flamingoes in the Andes to marine turtles in the Indian Ocean and off Southeast Asia.

Lowercase agreements tend to require action by their signatories that may or may not be obligatory, depending on the agreement. Agreements have been negotiated to protect gorillas in Central Africa, birds that migrate between Africa and Eurasia, European bats, and whales in European waters. Uppercase AGREEMENTS necessitate adequate a priori knowledge of migratory patterns and population status and require signatories to follow strict guidelines outlined in the CMS text. By the early years of the twenty-first century no AGREEMENTS had been negotiated.

The main criticism of CMS has been that its area of responsibility is too large for a single treaty, with most progress made by agreements negotiated under the convention but not by the main convention itself. It has also been criticized for fostering weak, nonenforceable agreements, but supporters of the convention note that the range of formalities with which agreements can be made encourages participation by states that perhaps would otherwise be hesitant to engage in conservation. The effectiveness of CMS is limited by the absence of several large states important to migratory species, including the United States, Canada, China, Russia, and Brazil, although several of these do participate in daughter agreements.

*Adam B. Smith*

### Further Reading

Cioc, Mark. *The Game of Conservation: International Treaties to Protect the World Migratory Animals.* Athens: Ohio University Press, 2009.

DeSombre, Elizabeth R. *Global Environmental Institutions.* New York: Routledge, 2006.

## Carter, Jimmy

CATEGORY: Preservation and wilderness issues
IDENTIFICATION: American politician who served as governor of Georgia and as president of the United States
BORN: October 1, 1924; Plains, Georgia
SIGNIFICANCE: During his political career, Carter made many decisions that demonstrated an environmentalist agenda.

Jimmy Carter was born into a family that owned a general store and farm in the small community of Plains, Georgia. He was appointed to the U.S. Naval Academy, graduating in 1946, and served in the Navy until his father's death in 1953, when he assumed his father's business responsibilities.

Carter expanded his family's businesses and successfully ran for local political office. He was elected to the Georgia Senate in 1962 and 1964. He was elected governor of Georgia in 1970 and served in the office from 1971 to 1975. Governor Carter reorganized the state government, consolidating many functions and putting all environmental agencies under the Department of Natural Resources. He rejected a U.S. Army Corps of Engineers plan to dam the Flint River, the last free-flowing large river in western Georgia. This was apparently the first such action ever taken by a governor of a U.S. state. He helped to arrange for the greater part of Cumberland Island to be given to the state of Georgia; it would later become Cumberland Island National Seashore. He also began acquiring land along the Chattahoochee River in Atlanta for state parks; these later became parts of the Chattahoochee River National Recreation Area.

Carter ran for the U.S. presidency in 1976 and won the election by a narrow margin over incumbent Gerald Ford. Early in his term, President Carter vetoed a public works bill that included nineteen water projects that he considered economically unjustified and environmentally unsound. Under intense political pressure he later signed a compromise bill that included nine of the projects. He also issued executive orders that directed federal agencies to protect or restore wetlands and floodplains wherever possible as a matter of government policy.

During the Carter administration a large number of important environmental acts were passed and signed into law by the president. These included the Alaska National Interest Lands Conservation Act (1980), the Fish and Wildlife Conservation Act (1980), and the Comprehensive Environmental Response, Compensation, and Liability Act, commonly referred to as Superfund (1980). It is remarkable that all of this was accomplished in the midst of a very difficult single term in office. President Carter's term began with a worldwide economic recession and instability in the international petroleum market, resulting in large national budget deficits, severe inflation, and high interest rates; it ended with the overthrow of the shah of Iran and the seizure of the U.S. embassy in Iran by the revolutionary Islamic government.

In response to the petroleum market problem, President Carter requested the creation of a cabinet-level Department of Energy. The mission of this department included research on reducing American dependence on fossil fuels and the development of alternative energy resources. Over the decades since, this research has had very desirable environmental effects.

*Robert E. Carver*

FURTHER READING

Horowitz, Daniel. *Jimmy Carter and the Energy Crisis of the 1970s: A Brief History with Documents*. Boston: Bedford/St. Martin's Press, 2005.

Stine, Jeffrey K. "Environmental Policy During the Carter Presidency." In *The Carter Presidency: Policy Choices in the Post-New Deal Era*, edited by Gary M. Fink and Hugh Davis Graham. Lawrence: University Press of Kansas, 1998.

# Conservation

CATEGORY: Resources and resource management
DEFINITION: Planned use of natural resources to benefit the maximum number of people for as long as possible
SIGNIFICANCE: The goals of conservation often conflict with the drive toward the exploitation of natural resources for immediate economic gain and also with the ethic of preservation, which promotes the indefinite preservation of natural resources in their undisturbed state.

For centuries few people recognized that nature's resources are finite. Only since the mid-nineteenth century has the issue of conserving resources been taken seriously. In North America, the eighteenth and nineteenth centuries were largely devoted to conquering the wilderness. As pioneers cleared forests, little or no thought was given to the idea that they could ever be exhausted. Bison and other animals were hunted nearly to extinction, and bounties were placed on wolves and other predators. In the seas, millions of seals were killed, and whales were hunted relentlessly.

Almost too late, it became apparent to many people that soils were being depleted by erosion and that once-plentiful native plant and wildlife populations were in decline. Even the quality of air, surface water, and groundwater deteriorated. An increasing number of people began calling for an end to environmentally destructive practices. An important influence was that of diplomat and naturalist George Perkins Marsh, whose travels had made him keenly aware of the results of centuries of land abuse in Europe. Marsh's book *Man and Nature: Or, Physical Geography as Modified by Human Action* (1864) defined the connections among soil, water, and vegetation. From such influences came the public park movement, including the establishment of Yellowstone National Park.

### EMERGENCE OF THE MODERN CONSERVATION MOVEMENT

During the early twentieth century, the living standards of most Americans generally improved, but at the same time it was recognized that consumption of natural resources had to be controlled. Out of this realization emerged a human-centered conservation philosophy known as wise use, which dictated that human beings should use nature in such a way that its resources could continue to be used over long periods of time.

Among the conservationists of the era were U.S. president Theodore Roosevelt and forester Gifford Pinchot. Roosevelt increased the extent of the national forests in the United States and created many wildlife refuges. Pinchot, a friend and associate of Roosevelt, became chief forester of the U.S. Department of Agriculture and greatly influenced the conservation movement. He applied European methods of managing forests that were consistent with the wise-use philosophy. This put him in conflict with naturalist John Muir and others who wished to preserve wilderness areas in their natural state. Two opposing camps developed: On one side were preservationists, who argued that nature deserves to be protected for its own sake; on the other were conservationists, who believed in regulated exploitation.

During the 1920's and 1930's, scientific advances and economic problems influenced conservation views and policies. The ecosystem concept, the principle that nature is composed of local units with interacting living and nonliving components, developed. It was to become the cornerstone of the science of ecology and an important tool of conservation. During this same period a new, scientific approach to wildlife management emerged. Aldo Leopold, regarded as the father of the new science, wrote the influential textbook *Game Management* (1933), although he is better known today for his collection of essays *A Sand County Almanac, and Sketches Here and There*, which was published in 1949, soon after Leopold's death. Paul B. Sears achieved prominence with *Deserts on the March* (1935), which vividly dramatized the problems of the Dust Bowl of the 1930's.

The conservation legacy of President Franklin D. Roosevelt includes his attempts to restore the depressed U.S. economy by creating various work programs, including the Civilian Conservation Corps (CCC). Many of these programs were aimed at soil conservation, reforestation, and flood control.

### POSTWAR DEVELOPMENTS

In the two decades following World War II, concerns about the Cold War and economic expansion kept conservation far from the forefront of American thought. As the population rapidly expanded, air and water pollution worsened, grasslands became overgrazed, and agricultural chemicals were used in increasing quantities. The complacency of the times was shattered by the publication of Rachel Carson's *Silent Spring* (1962), which warned that dichloro-diphenyl-trichloroethane (DDT) and other pesticides were threatening the lives of animals and humans. The book caused widespread public concern, and laws were ultimately passed that outlawed DDT in the United States.

Under President John F. Kennedy's administration, attention was given once more to environmental issues in general and to conservation. Funds were appropriated to improve air quality, and new land was acquired for parks. Stewart Udall, secretary of the interior under Kennedy and his successor, Lyndon B. Johnson, advocated a new positive attitude toward protection of the environment. During the 1960's, a time of great scientific activity and social ferment, ecologists employed advanced technologies in conducting ecosystem analyses. Among the social movements of the decade was a new environmentalism, which culminated in the first Earth Day celebration in 1970.

The 1970's saw the expansion of environmentalism, and the term "environmentalism" came to be understood as almost synonymous with "conservation." The former clear distinction between the terms "conservation" and "preservation" was blurred. Environmental organizations continued to elaborate their views and increase their visibility. More action-oriented alternative groups used sabotage tactics to stop development in wilderness areas. In contrast, the Nature Conservancy became known for its businesslike policy of acquisition, protection, and management of natural areas. In response to the growing awareness of environmental issues among Americans, a new round of federal legislation was passed. Under President Richard Nixon, the Environmental Protection Agency (EPA) was created in 1970. The revised Clean Air Act of 1970 was followed by the Clean Water Act of 1972, and in 1973 Congress enacted the Endangered Species Act.

### POLICY AND POLITICS

The 1980's were marked by new conservation problems and a general indifference at the federal level in the United States. Among the global problems identified were stratospheric ozone depletion and global warming. Reversals of environmental protections occurred during the administration of President Ronald Reagan after he appointed James Watt as secretary of the interior. Watt attempted to dismantle many of the environmental protection programs that had been put in place during the previous decades.

The 1990's had their own environmental challenges and successes. After twelve years of apathy, conservationists were reinvigorated by the election of Bill Clinton as president and high-profile environmentalist Al Gore as vice president. Despite continued optimism among environmentalists and significant gains in environmental protections, however, an anti-environmental movement also developed. Many politically conservative individuals and organizations began to assert that environmentalism had gone too far. Business leaders and property owners complained of overregulation, and even some religious authorities expressed concerns that nature had been elevated in importance above humans. Nevertheless, the majority of Americans continued to feel that nature deserves to be protected. The late twentieth century saw a growing awareness of the importance of biodiversity and increasing support for the idea that all species should be protected against extinction. From this trend emerged the value-laden science of conservation biology.

Under the presidency of George W. Bush, the early twenty-first century was marked by business- and industry-friendly policies that diminished protections for wildlife and the environment. However, in the final month of his administration, Bush took the historic action of designating three areas of the Pacific Ocean as marine national monuments. This status protects the Marianas Trench, Pacific Remote Islands, and Rose Atoll marine national monuments from destruction or extraction of their resources, waste dumping, and commercial fishing.

In April, 2010, President Barack Obama held a national conservation conference that echoed one convened by President Theodore Roosevelt more than a century earlier. Dubbed America's Great Outdoors Initiative, Obama's conference took into account the developed, urban, and populous nature of the country. The twenty-first century strategy for the nation that came out of the conference proposed partnerships among government, nonprofit organizations, and private landowners for land preservation efforts that would be smaller in scale than national parks and designated wilderness areas and thus less likely to encounter heated political opposition.

*Thomas E. Hemmerly*
*Updated by Karen N. Kähler*

FURTHER READING

Adams, Jonathan S. *The Future of the Wild: Radical Conservation for a Crowded World.* Boston: Beacon Press, 2006.

Carroll, Scott P., and Charles W. Fox, eds. *Conservation Biology: Evolution in Action.* New York: Oxford University Press, 2008.

Chiras, Daniel D., and John P. Reganold. *Natural Resource Conservation: Management for a Sustainable Future.* 10th ed. Upper Saddle River, N.J.: Benjamin Cummings/Pearson, 2010.

Cox, George W. *Conservation Biology: Concepts and Applications.* 2d ed. Dubuque, Iowa: Wm. C. Brown, 1997.

Groom, Martha J., Gary K. Meffe, and Carl Ronald Carroll. *Principles of Conservation Biology.* 3d ed. Sunderland, Mass.: Sinauer, 2006.

Hambler, Clive. *Conservation.* New York: Cambridge University Press, 2004.

Kline, Benjamin. *First Along the River: A Brief History of the U.S. Environmental Movement.* 3d ed. Lanham, Md.: Rowman & Littlefield, 2007.

Macdonald, David W., and Katrina Service, eds. *Key Topics in Conservation Biology.* Malden, Mass.: Blackwell, 2008.

Pullin, Andrew S. *Conservation Biology.* New York: Cambridge University Press, 2002.

# Conservation movement

CATEGORIES: Resources and resource management; preservation and wilderness issues

IDENTIFICATION: Activities of individuals, private organizations, and government agencies to preserve natural resources or establish policies for using them wisely

SIGNIFICANCE: The early conservation movement focused attention on the perilous state of natural resources as human societies moved toward increasing commercialization and urbanization in the nineteenth and twentieth centuries. Proponents of conservation continue to work to convey the importance of resource sustainability.

The roots of the conservation movement can be traced to Europe during the late eighteenth century. Individuals repulsed by the growing commercialization and dehumanization brought on the by the Industrial Revolution promoted a return to nature. While some areas had been preserved in their natural state for centuries, either as hunting grounds or as refuges for the upper classes, during the nineteenth century governments in Europe and in North America began setting aside green space, even in urban areas, for recreation and relaxation. In some countries, notably Germany, government control over forests and rivers was used not only to conserve natural resources but also to generate revenues from the harvest of renewable resources such as trees and wildlife. Coordinated efforts to preserve wilderness areas, however, emerged first in the United States. The initiatives of individuals and organizations involved in these efforts grew into the modern conservation movement.

BEGINNINGS OF THE MOVEMENT

From its inception, the conservation movement in the United States was highly decentralized, but people who became involved shared an overarching goal: to preserve some aspect of the natural world from the ravages of rampant commercial development and urbanization. Efforts to save some of America's wilderness areas from the encroachment of civilization led to the creation of Yellowstone National Park in 1872, and some states had taken steps to set aside particularly beautiful areas (many of them difficult to reach). On the other hand, the first true conservation bill passed by the U.S. Congress in 1876 was vetoed by President Ulysses Grant.

Although American writers such as Henry David Thoreau had celebrated the value of the natural landscape, the first true conservationist to emerge on the national scene was John Muir. A Scotsman who had settled in California, Muir spent most of his adult life writing about the beauties of the wilderness, particularly the Yosemite Valley, a preserve set aside by the state in 1864. At that time many people looked upon the wilderness as a place of escape from the grime of city life. Powerful commercial interests, however, saw unsettled areas as opportunities for profit. The mining and timber industries coveted the natural resources there, farmers and ranchers sought lands for grazing, and railroads saw possibilities in creating tourist destinations or shorter routes between population centers.

Muir was one of dozens of amateurs who made their livings by other means but devoted their lives to saving the country's wilderness for future generations. Many were responsible for creating new national parks or monuments, saving sites that had been commercialized (such as Niagara Falls), or establishing wildlife sanctuaries to preserve dwindling numbers of birds, fish, and mammals. Muir lobbied to preserve Yosemite and additional wilderness sites. Eventually he found a strong ally in President Theodore Roosevelt, a lifelong sportsman and conservationist who used his political power to establish numerous national monuments to save lands from mining, timbering, and settlement. Muir was a preservationist. His goal was to have lands set aside, protected from all development or use.

A second important figure in the conservation movement, Gifford Pinchot, could best be described as a utilitarian. Pinchot earned a degree in forestry and eventually became head of the U.S. Forest Service, making him one of the first professional conservationists. He argued that natural resources should be managed carefully, but renewable resources should be available for commercial purposes. These two views came to dominate the debate among conservationists for nearly a century as groups struggling to preserve sites intact clashed with those who saw responsible use as a viable alternative to saving lands from outright, wholesale exploitation.

Over the next century the conservation movement would be advanced by both amateurs and professionals. A few crossed from one group to the other. Notable among them was Stephen T. Mather, who made a fortune in business and then turned his talents to creating the National Park Service in 1916, for which he served as the first director. Intent on preserving the national parks for future generations, Mather often found himself pitted against Pinchot and the Forest Service in debates concerning the proper balance between preservation and commercial use of national lands.

Spurred by Muir's leadership, in 1892 a group of Californians formed the Sierra Club, which would become a driving force in promoting conservation throughout the United States. Similar organizations sprang up in the East, many modeled on the first such club, the Appalachian Mountain Club, established in 1876.

Eventually the idea of conservation grew to include wildlife as well as landscape. Ironically, some of the most avid conservationists were sportsmen. They recognized that careful management of game populations would ensure that hunters could enjoy their sport for decades to come. Aware of what wanton slaughter had done to the American bison population during the nineteenth century, they often lobbied for limits on the numbers of animals that individual hunters could kill.

Conservationist groups were not always successful, however. Despite the protests of the Sierra Club and others, the U.S. government agreed to the creation of the Hetch Hetchy Dam in the Yosemite area to provide water for San Francisco, causing the loss of millions of hectares of natural landscape. Lobbying did push several presidents to create wildlife refuges, setting aside lands where game animals lived or that they used as migratory routes. Eventually these sanctuaries became part of the National Wildlife Refuge System.

## Between the World Wars

By the end of World War I a number of conservation groups were active in the United States. Individual organizations, however, tended to focus on single aspects of conservation. For example, the Audubon Society, founded in 1905, limited its interest to the plight of birds, while the Izaak Walton League, established in 1922, was concerned with the condition of rivers and lakes because its members were involved in sportfishing. In fact, during this period many conservation organizations were underwritten by firearms companies and sports businesses. As a result, the utilitarian view of conservation tended to be favored in political decisions.

Concurrently, preservationists such as the National Park Service's Mather tried to have more lands set aside; on occasion these efforts were supported by financiers such as John D. Rockefeller, Jr., who purchased properties to expand or create parks. Some narrow-minded conservation efforts led to ecological disasters, however. For example, by the 1920's the gray wolf population in Yellowstone National Park was eliminated, ostensibly to make the park safer for tourists (and for the livestock of nearby ranchers). The absence of these predators led to an explosion of the ruminant population, which then overgrazed the available vegetation; eventually thousands of elk and deer starved.

The election of Franklin D. Roosevelt as president in 1932 brought one of the most active proponents of the conservationist movement into the White House. Roosevelt had a genuine love of the outdoors and was a conservationist at heart. He was a utilitarian when it came to conservation, however, and above all a politician. He believed that a government-controlled program of land retirement, soil and forest restoration, flood control, and hydroelectric power generation would serve two aims: Natural landscapes could be preserved or put to good use at the same time people could be given meaningful employment.

Combining his zeal for conservation with the need to revitalize the American economy during the Great Depression, Roosevelt created numerous programs to further the improvement of the nation's natural resources. Notable among these were the Civilian Conservation Corps and the Works Progress Administration, agencies that put people to work improving national parks and other wilderness areas, making them accessible to the public, and restoring many of the renewable resources that had been lost over the years to commercial development. Among his principal successes, Roosevelt oversaw the opening of Great Smoky Mountains National Park in 1934.

Not all of the projects Roosevelt promoted under the banner of conservation were beneficial to the natural environment, however. The Tennessee Valley Authority oversaw the erection of numerous dams to provide hydroelectric power to rural communities, in the process destroying millions of hectares of forestlands and valleys. Politically, the most significant change in the conservation movement under Roosevelt was the affiliation of environmental causes with the Democrats rather than the Republicans, who had been leaders in conservation in previous years.

## Conservation after World War II

After World War II ended, the United States entered a period of economic prosperity that saw the population grow and industry expand to meet worldwide demand for American goods. As a result, individuals and corporations interested in development lobbied for relaxation of conservation rules that had been put in place during the previous six decades. Many argued for a return to commercial operations in government-protected areas. Conservation groups, which by then numbered in the dozens, fought back with mixed success to maintain and expand the protections in place.

Organizations such as the Sierra Club and the Wilderness Society developed activist tactics to bring attention to the long-term dangers of development in wilderness areas. These groups were successful in defeating a plan to create a dam on the Green River at Echo Park, on the Utah-Colorado border, during the 1950's. Most notably, under the leadership of the Wilderness Society's executive secretary Howard Zahniser, conservationists prodded Congress to pass the Wilderness Act in 1964. That law set aside more than 3.6 million hectares (9 million acres) of land, restricting even minimal development or human use.

The publication of Rachel Carson's *Silent Spring* in 1962, detailing the environmental dangers of toxic waste, gave new impetus to the conservationists and sparked the larger environmental movement. As a consequence, during the 1960's conservation groups began to unite with those interested in larger environmental issues. Sierra Club executive director David Brower became a leading spokesman for environmental causes. The most significant success achieved by these groups in lobbying governmental agencies was the passage of the National Environmental Policy Act of 1969, which established the government's role in protecting humans and the natural environment from damage caused by human development or technology. In December, 1970, the Environmental Protection Agency was established to enforce environmental policies throughout the United States.

## International Conservation Efforts

By the middle of the twentieth century, the modern conservation movement had become international. The record of conservation groups in Europe was strong; many had been successful in continuing and strengthening already existing laws to preserve natural resources. Beginning during the 1960's conservationists in Australia led highly organized campaigns to reverse the effects of human contamination on the natural environment, specifically targeting problems with the quality of soils and the eradication of several species of animals. In Africa, Asia, and South America, however, conservationists had mixed success. Often conservation efforts conflicted with the needs of the people for resources to support life. This was especially true in Africa, where native peoples relied on wildlife for sustenance, and in South America, where lands covered by rain forests were needed for farming. Some successes were achieved when native populations were convinced that conservation would provide long-term economic stability.

At the same time, however, commercial interests continued to harvest animals and timber far in excess of the population's needs, ignoring the environmental catastrophes caused by their actions. Often, government-sponsored programs wreaked havoc on natural resources. Nowhere was this more apparent than in China, where the need for energy to fuel the country's growing population and expanding industrial base led to construction of the massive Three Gorges Dam on the Yangtze River, flooding millions of hectares, destroying or seriously affecting land and riverine populations. From 1970 onward, however, China began to take an active approach to wildlife preservation, as did several other Asian nations.

One of the most successful conservation initiatives undertaken outside the United States has been that of the Central American nation of Costa Rica. Beginning in 1970 the Costa Rican government worked to create a highly developed national park system where conservation practices would be strictly enforced. By 2000 nearly 25 percent of Costa Rica's landmass had been set aside for preservation.

*Laurence W. Mazzeno*

FURTHER READING

Chester, Charles A. *Conservation Across Borders: Biodiversity in an Interdependent World.* Washington, D.C.: Island Press, 2006.

Coggins, Chris. *The Tiger and the Pangolin: Nature, Culture, and Conservation in China.* Honolulu: University of Hawaii Press, 2003.

Evans, Sterling. *The Green Republic: A Conservation History of Costa Rica.* Austin: University of Texas Press, 1999.

Fox, Stephen. *John Muir and His Legacy: The American Conservation Movement.* Boston: Little, Brown, 1981.

Gibson, Clark L. *Politicians and Poachers: The Political Economy and Wildlife Policy in Africa.* New York: Cambridge University Press, 1999.

Maher, Neil M. *Nature's New Deal: The Civilian Conservation Corps and the Roots of the American Environmental Movement.* New York: Oxford University Press, 2008.

Markus, Nicola. *On Our Watch: The Race to Save Australia's Environment.* Melbourne: University of Melbourne Press, 2009.

Minteer, Ben and Robert Manning, eds. *Reconstructing Conservation: Finding Common Ground.* Washington, D.C.: Island Press, 2003.

Phillips, Sarah T. *This Land, This Nation: Conservation, Rural America, and the New Deal.* New York: Cambridge University Press, 2007.

Runte, Alfred. *National Parks: The American Experience.* 4th ed. Lanham, Md.: Taylor Trade, 2010.

## Conservation Reserve Program

CATEGORIES: Agriculture and food; land and land use

IDENTIFICATION: Cropland reduction and conservation program of the U.S. Department of Agriculture

DATE: Created on December 23, 1985

SIGNIFICANCE: Initially established to reduce agricultural surpluses by encouraging farmers to reduce the amounts of land they devoted to crops, the Conservation Reserve Program has grown to become an environmental protection program as well, preventing soil erosion and reducing carbon in the atmosphere.

The history of agriculture in the United States reveals a pattern of crop production increasing faster than demand, resulting in downward pressure on crop prices and farm income. Improvements in equipment, chemicals, and seed genetics have increased yields. In 1982, farmers produced a surplus of 230 million metric tons (254 million U.S. tons) of crops, swamping the available storage capacity. The government response, announced by President Ronald Reagan on January 11, 1983, was a payment-in-kind (PIK) program. Under this program, farmers were given certificates for crops in storage in return for taking land out of production.

The PIK program was generous, with wheat farmers getting up to 95 percent of the normal crop. Farmers responded by removing 33.3 million hectares (82.3 million acres) of cropland from the 1983 growing season. This steep reduction from the 170 million hectares (420 million acres) planted in 1982 had a devastating economic impact on small towns dependent on agriculture. Farmers bought less fuel and fertilizer, and they did not hire seasonal workers. The resulting stress on rural communities made it clear to policy makers that wide swings in the amounts of land planted were unacceptable. The U.S. Congress responded to the overproduction issue with the Conservation Reserve Program (CRP), which was created under Title XII of the 1985 Food Security Act.

The twin goals of CRP were to decrease crop surpluses through reduction in land area planted and to prevent cropland erosion. Enrollment was voluntary, with farmers submitting bids for the amount of payment required to remove land from production for ten to fifteen years. Participants were paid a rental fee plus half the cost of establishing a permanent cover of trees or grasses. Enrollment increased rapidly, with approximately 13.8 million hectares (34 million acres) enrolled during the first nine sign-up periods from 1986 through 1989.

Over time Congress enhanced the environmental protection aspects of CRP. The Food, Agriculture, Conservation, and Trade Act of 1990 extended the enrollment period through 1995 and added water-quality protection as a criterion for land selected. The Federal Agriculture Improvement and Reform Act of 1996 included an Environmental Benefits Index for selection of land suitable for enrollment.

The environmental impact of CRP is substantial—ground-cover plants on CRP lands annually remove an estimated 17 million metric tons (18.7 million U.S. tons) of carbon from the atmosphere and reduce soil erosion by more than 443 million metric tons (488 million U.S. tons). These lands also provide critical habitat for wildlife, including upland game birds, grassland songbirds, and prairie mammals. CRP lands are particularly important to migratory waterfowl. Duck species that had historically nested in the Prairie Pothole region of the northern plains were in serious trouble during the 1980's, with populations near their lowest level in the preceding fifty years. Habitat provided by CRP increased the ducks' nesting success, helping the population rebound from 25.6 million breeding ducks in 1985 to 42 million in 2009.

CRP continues as an important environmental and agricultural program. The Food, Conservation, and Energy Act of 2008 extended CRP enrollment authority to September 30, 2012. It set the enrollment authority at 15.8 million hectares (39 million acres) through 2009 but reduced enrollments to 13 million hectares (32 million acres) for fiscal years 2010 to 2012. In 2009 there were 14 million hectares (34.7 million acres) enrolled at an annual rental cost of $1.76 billion per year, an average cost of about $125 per hectare ($50 per acre) per year. With considerable support for CRP from agricultural producers, organizations concerned with waterfowl, and the environmental community, the program is likely to continue to be included in future farm legislation.

*Allan Jenkins*

FURTHER READING

Hamilton, James T. *Conserving Data in the Conservation Reserve.* Washington, D.C.: RFF Press, 2010.

Napier, Ted, Silvana M. Napier, and Jiri Tvrdon, eds. *Soil and Water Conservation Policies and Programs: Successes and Failures.* Boca Raton, Fla.: CRC Press, 2000.

# Controlled burning

CATEGORIES: Resources and resource management; forests and plants; agriculture and food

DEFINITION: Intentional setting of a fire to accomplish a specific purpose

SIGNIFICANCE: The controlled use of fire can benefit forests by preventing larger, uncontrollable wildfires and by removing undesirable plants, promoting certain animal species, and stimulating the germination of some seeds; controlled burns are also used as a means of fighting wildfires. The controlled burning used in agriculture, however, is often accompanied by negative environmental impacts such as air pollution and eventual soil depletion.

Controlled burning is one tool that foresters use to help contain wildfires. There are three ways to extinguish a fire: Cool the fuel below its kindling temperature, deprive the fire of oxygen, or deprive the fire of fuel. Pouring water on a fire addresses the first two of these; the setting of a controlled burn known as a backfire is one way to accomplish the third. If the wind is blowing toward a raging wildfire, firefighters may set a line of fire in front of the wildfire. When correctly controlled, the backfire consumes the available fuel before the wildfire can reach that area; the wildfire then dies out for lack of fuel.

Controlled burning is also used as a way of preventing large, unmanageable wildfires. In forests and other wilderness areas, lightning strikes start many fires every year, and nature has adapted to relatively frequent burning. Fires promote new growth and renewal of a forest in several ways. For example, cones from the giant sequoia and the lodgepole pine do not release their seeds unless they have been dried or heated, and heating is more efficient. Fires clear the ground of brush, leaves, pine needles, and dead wood, allowing seeds to sprout in the ground, which is moist and enriched by nutrients leached from the ashes. (Forest floor litter dries out more quickly than the soil underneath, and seedlings that sprout in the litter generally die.) In addition, fire burns away plants that would have blocked sunlight from seedlings.

If too much time elapses between fires, however, fuel can accumulate to dangerous levels on the forest floor. When a fire starts after a long period of fuel accumulation, it is likely to burn so hot that standing trees will be consumed, and in a strong wind, flames will jump from the crown of one tree to the crown of the next. Crown fires race quickly through a forest and are very difficult to extinguish. Foresters have come to understand that it can be far less destructive to a forest to use controlled burns to keep the fuel load at a safe level.

Other uses of controlled burning are found in agriculture. After the last ice age, about 11,000 years ago, hunter-gatherers settled down to farming and adopted the slash-and-burn technique for clearing land. Large trees were cut down, and smaller trees and underbrush were burned; this both cleared the land and released nutrients into the soil. After a few years, when the land began producing less, a new area was selected and the cycle repeated. This practice continues in some parts of the world in the twenty-first century, notably in the Amazon rain forest and in Southeast Asia. In the United States, some 10 percent of Oregon grass-seed farmers use fire to kill competing weeds, weed seeds, insects, and rodents because burning is cheaper than other methods. Oklahoma wheat farmers burn wheat stubble for similar reasons.

Range managers in the Wichita Mountains Wildlife Refuge have adopted the controlled burning practice known as patch burning. They set controlled fires in areas of rangeland from 40 hectares (100 acres) to 8,100 hectares (20,000 acres) in size every few years to provide the bison in the refuge with the nutritious, tender grass shoots that grow in the burned areas. Fuel is allowed to accumulate in unburned areas, which are then burned to repeat the cycle.

*Charles W. Rogers*

FURTHER READING

Carle, David. *Introduction to Fire in California.* Berkeley: University of California Press, 2008.

Reinhart, Karen Wildung. *Yellowstone's Rebirth by Fire: Rising from the Ashes of the 1988 Wildfires.* Helena, Mont.: Farcountry Press, 2008.

Ribe, Tom. *Inferno by Committee: A History of the Cerro Grande (Los Alamos) Fire, America's Worst Prescribed Fire Disaster.* Victoria, B.C.: Trafford, 2010.

## Convention Relative to the Preservation of Fauna and Flora in Their Natural State

CATEGORIES: Treaties, laws, and court cases; preservation and wilderness issues; animals and endangered species

THE CONVENTION: International agreement that established preservation policies for European colonies in Africa

DATE: Opened for signature on December 8, 1933

SIGNIFICANCE: The Convention Relative to the Preservation of Fauna and Flora in Their Natural State was among the first international agreements concerned with issues of conservation, although the signatory nations were concerned primarily with protecting animals from extinction so that they would remain available for game hunting.

In 1900 the European nations that had recently divided sub-Saharan Africa among themselves and established colonial governments there signed the first international conservation treaty, the Convention for the Preservation of Animals, Birds, and Fish in Africa. The men who drafted this document did not recognize any inherent value in living creatures—they were not protecting animals because they felt animals had a right to live. The intention of the treaty was to preserve the populations of animals that were popular trophies for hunters, such as elephants and giraffes, and encourage the eradication of animals harmful to agriculture, including lions, leopards, and wild dogs.

In 1930 a surveying expedition sponsored by the British Society for the Protection of the Fauna of the Empire made it clear that the 1900 treaty was ineffective from a conservation standpoint. Elephants and other animals were still being overhunted, and several animal and plant species were drawing closer to extinction. It was proposed that an expanded system of national parks be established in East and Central Africa to protect species without substantially limiting human activity. The national parks would be under the control of the colonial governments. The public would be encouraged to visit the national parks to observe the plants and animals, but no "hunting, killing, or capturing" would be permitted within park boundaries.

Several nations that held substantial amounts of land in Africa met in London in 1933 to discuss these issues. The resulting Convention Relative to the Preservation of Fauna and Flora in Their Natural State was signed by South Africa, Belgium, the United Kingdom, Egypt, Spain, France, Italy, Portugal, and the Sudan. It established national parks for public enjoyment and "strict natural reserves" for the exclusive use of scientists. One plant species and twenty animals—including gorillas, white rhinoceroses, and shoebill storks—were fully protected by the treaty, which entered into force on January 14, 1936. New rules for hunters outside the parks forbade the use of cars and aircraft to chase or herd animals and also prohibited poison and traps.

However, neither the treaty nor the discussions leading up to it considered the role black Africans might play in preserving or endangering the fauna and flora. The treaty was made by Europeans to ensure that white people would have enough animals to hunt. Much of the land newly dedicated to national parks had been home to Africans who were now forbidden to hunt, farm, or live on that land. Animals were protected or not protected according to their usefulness for or danger to white hunters and settlers, without consideration of which animals provided food or presented a danger to native villagers.

*Cynthia A. Bily*

FURTHER READING

Louka, Elli. *International Environmental Law: Fairness, Effectiveness, and World Order.* New York : Cambridge University Press, 2006.

Suich, Helen, and Brian Child, with Anna Spenceley, eds. *Evolution and Innovation in Wildlife Conservation: Parks and Game Ranches to Transfrontier Conservation Areas.* Sterling, Va.: Earthscan, 2009.

## Cross-Florida Barge Canal

CATEGORY: Preservation and wilderness issues
IDENTIFICATION: Intended shortcut for shipping across Florida that would have linked the Gulf of Mexico with the Atlantic Ocean
SIGNIFICANCE: Efforts to build the Cross-Florida Barge Canal caused decades of controversy among environmentalists, government officials, and business interests until it was finally decommissioned in 1990 and the land was converted into a state recreation and conservation area.

Florida's unique peninsular shape and its 6,115 kilometers (3,800 miles) of tidal shoreline have long frustrated military, industry, and shipping interests. Since no river cuts across the state, ships and barges have had to travel around Florida's southern tip. In earlier times, this trip was often a perilous undertaking because of turbulent storms and the existence of many dangerous reefs and shoals.

Even before Florida became a state, various interest groups and individuals began calling for the creation of a transpeninsula waterway. Bowing to pressure, the Florida legislature created the Florida State Canal Commission in 1821 to explore the possibility of building such a canal. Five years later, the U.S. Congress adopted the cause and authorized the first of twenty-eight surveys that were carried out to find a convenient and safe route for an inland ship canal across the state. One proposal after another either proved to be impractical or failed to gain political support.

CONSTRUCTION HISTORY

In 1935 President Franklin D. Roosevelt, intending to ease unemployment problems in Florida, used $5 million in federal relief money as start-up funds for the construction of a ship canal. The proposal called for a 9-meter (30-foot) sea-level ship canal that would stretch across the north-central part of the state from the Atlantic Ocean to the Gulf of Mexico. On September 19, 1935, Roosevelt pressed a telegraph key from his office in Washington, D.C., and set off an explosive blast in Florida that officially began the canal's construction.

Roosevelt's action received popular support across north-central Florida, a region where many communities welcomed the new jobs and economic boost the canal was expected to generate. Intense opposition to the canal also existed, however. Railroad interests, fearing competition from shipping interests, hotly objected to the canal construction and lobbied in Congress to end it. Many south Floridians, fearing that a ship canal would allow saltwater intrusion to jeopardize the state's downstream supply of underground water, joined the protest. Three years after the project began, Roosevelt yielded to political pressure and cut off funding; he called on Congress to help, but Congress failed to appropriate any money for the canal, and construction ground to a halt.

World War II rekindled interest in the canal. German submarine attacks against U.S. ships along the Florida coast prompted Congress to ask the Army Corps of Engineers to reexamine the issue of building a canal to meet the nation's wartime needs. The corps responded with scaled-back plans for a barge canal, but Congress again failed to provide funding, and the project stagnated for years. In 1964 Congress finally authorized $1 million to get construction under way and promised more funds later. The Army Corps of Engineers now had responsibility for the project. Its engineers came up with plans for a five-lock waterway that would stretch 296.8 kilometers (184.4 miles) from Port Inglis on Florida's west coast to the Intracoastal Waterway at the St. Johns River on the east.

### Opposition from Environmentalists

Opposition to the project quickly developed. Environmentalists argued that the canal would disrupt the natural flow of rivers and creeks in the region, flood woodland areas, and destroy many endangered and threatened plants and animals. In 1969 the Environmental Defense Fund (EDF), along with the Florida Defenders of the Environment, sued in a U.S. district court to stop construction. Nearly two years later, on January 15, 1971, the court granted the plaintiffs an injunction. Four days later, President Richard Nixon, citing environmental and economic concerns, issued a presidential order that suspended construction. By

> ### Protecting Her Own Backyard
>
> *According to the Florida Defenders of the Environment, the organization she founded to fight construction of the Cross-Florida Barge Canal, Marjorie Harris Carr had this to say when asked why she took on the cause:*
>
> Why fight for the Ocklawaha River? The first time I went up the Ocklawaha, I thought it was dreamlike. It was a canopy river. It was spring-fed and swift. I was concerned about the environment worldwide. What could I do about the African plains? What could I do about India? How could I affect things in Alaska or the Grand Canyon? But here, by God, was a piece of Florida. A lovely natural area, right in my backyard, that was being threatened for no good reason.

now, $74 million had been spent to build less than one-third of the canal. Great stretches of trees had been leveled, rivers and streams altered, two locks built, a dam constructed, and much earth moved.

In March, 1974, the Middle District Court of Florida overruled Nixon's action, but it upheld the injunction. Supporters of the canal received another blow in 1977 when both the Army Corps of Engineers and the Florida cabinet went on record calling for an end to construction. Despite these actions, canal proponents continued to lobby in Congress and succeeded in postponing a complete dismantling of the transwaterway project for years. Finally, in 1990, both the U.S. Congress and the Florida legislature officially and permanently deauthorized the barge canal.

Environmentalists hailed the defeat of the Cross-Florida Barge Canal as a major environmental victory, but they soon faced new problems. Debates arose over questions of what was to be done with completed sections of canal and the adjacent canal lands. In 1993 the Florida legislature resolved the issue when it authorized the conversion of 177 kilometers (110 miles) of the defunct canal zone into a huge nature preserve named the Cross-Florida Greenway State Recreation Area, or the Cross-Florida Greenway (renamed the Marjorie Harris Carr Cross-Florida Greenway in 1998, in honor of a leader of the movement to stop the canal). In addition to providing land that humans can use for outdoor recreation, the 28,300-hectare (70,000-acre) corridor also serves as a permanent wildlife refuge—one of the largest in the southern United States.

One controversial issue remained unresolved. It focused on the fate of the Rodman Dam, which was built in the center of the state on the Ocklawaha River prior to the final decommissioning of the canal. Environmentalists argued that since the dam was no longer needed, it should be demolished so that the Ocklawaha River could be restored to its natural flow pattern. Supporters of the dam—including local merchants and the bass fishing enthusiasts and fish camp owners who used the reservoir created by the dam—wanted to keep it in place, citing the reservoir's economic benefits to the local community as a recreational area. Even though Florida governor Lawton Chiles, the state cabinet, and the Department of Environmental Protection expressed support for restoration efforts of the Ocklawaha River and elimination of the Rodman Dam, the state legislature withheld the necessary funds.

*John M. Dunn*

FURTHER READING

Buker, George E. *Sun, Sand, and Water: A History of the Jacksonville District U.S. Army Corps of Engineers, 1821-1975.* Jacksonville, Fla.: U.S. Army Corps of Engineers, 1980.

Flippen, J. Brooks. *Nixon and the Environment.* Albuquerque: University of New Mexico Press, 2000.

Florida Defenders of the Environment. *Environmental Impact of the Cross-Florida Barge Canal with Special Emphasis on the Ocklawaha Regional Ecosystem.* Gainesville: Author, 1970.

Irby, Lee. "A Passion for Wild Things: Marjorie Harris Carr and the Fight to Free a River." In *Making Waves: Female Activists in Twentieth-Century Florida,* edited by Jack E. Davis and Kari Frederickson. Gainesville: University Press of Florida, 2003.

Tebeau, Charlton W. *A History of Florida.* Miami: University of Miami Press, 1981.

## Debt-for-nature swaps

CATEGORY: Preservation and wilderness issues
DEFINITION: Strategy for reducing foreign debt in developing nations by trading debt forgiveness or debt reduction for guarantees of environmental activities by debtor nations
SIGNIFICANCE: Although some critics view debt-for-nature swaps as a form of cultural imperialism, these arrangements offer a way for environmental organizations to encourage developing nations to preserve ecosystems, conserve natural resources, and protect significant cultural sites.

An international finance crisis began during the late 1980's when many developing nations found that they had borrowed more from international lending institutions, mostly private banks, than they could repay. To recover some of the principal on the loans, banks began to sell the loans in financial markets, usually discounted to a fraction of their original value because of the threat of default. Several options to relieve this debt burden on developing nations were explored. One option was to refinance the debt to lower the interest rates and extend the time for repayment. Another strategy was to encourage domestic financial reforms in debtor nations by increasing domestic investment, expanding the domestic economy, raising taxes, reducing non-debt-related expenditures, or inflating the currency in order to pay the debt eventually. Finally, creditor nations and institutions could partially forgive the debt. Debt-for-nature swaps combine elements of all three of these options.

### How Debt-for-Nature Swaps Work

In debt-for-nature swaps, environmental organizations buy discounted debt in the financial markets from the banks. Instead of collecting the full amount of interest and principal from the debtor nations, the environmental organizations agree to forgive all or a portion of the debt if the debtor nations invest amounts up to the principal value of the debt in local preservation efforts, often the purchase of land for national parks or investment in skilled staff and improvements for existing parks. In these arrangements, conservation organizations benefit through an increase in local funding for conservation, and banks benefit because they have a new market for the debt. The debtor nations benefit in several ways: They are able to invest their funds in their own nations rather than transferring funds to the lending nations; they use inflated local currency rather than high-value, scarce, dollar-based foreign exchange; and they reduce their outstanding debt and interest payments on that debt, which allows them to continue making payments on the remaining debt and maintain their international credit ratings.

Debt-for-nature swaps erase the "debt overhang," that portion of debt that, if forgiven, allows the remaining debt to continue to be assumed by creditor nations or institutions with satisfactory levels of burden on the debtor nations and satisfactory debt payment risks for the creditor nations. Because environmental organizations purchase the debt at a discount, they are able to multiply their impacts on the environment. For example, in the first debt-for-nature swap in 1987, Conservation International purchased $650,000 in Bolivian debt for $100,000, then required the Bolivian government to establish a $250,000 endowment fund in local Bolivian currency to pay operating costs for a biosphere reserve in the Bolivian Amazon before erasing the debt.

In debt-for-nature swaps, debtor nations buy back a portion of their outstanding debt with an investment in a portion of their own natural capital. Natural capital includes caches of nonrenewable resources such as mineral or oil reserves, natural resources such as old-growth forests or endangered species habitat, historical artifacts such as prehistoric ruins or fossil deposits, and cultural resources such as the homelands of primitive indigenous peoples or significant architectural sites. The preservation of this natural capital has positive benefits for the ecology, sustainable development, and biodiversity. A swap is also likely to improve the debtor nation's economic ability to repay the remaining debt, because the preserved cultural and natural resources often serve as tourist attractions, cultural centers, and locations for academic and commercial research.

### Participation of Financial Institutions

Because some private lending institutions will not voluntarily participate in debt-relief processes that reduce the financial institutions' capital or potential profit from loans, governments in creditor nations either mandate financial institutions' participation or provide financial incentives, such as tax deductions and tax credits, to encourage participation. Creditor governments justify these actions as a component of their foreign economic development programs, as a component of their environmental programs, as philanthropic support for preservation of the earth's cultural and natural heritage, or as economically justifiable domestic self-interest. For example, encouraging nations in tropical zones to protect rain forests helps to reassure nations in the northern temperate zones that existing climate patterns supported by those rain forests will be maintained. Maintaining these climate patterns is necessary to prevent natural and human-made disasters, reduce the demands on industrialized nations to provide humanitarian relief from increased numbers of natural disasters, ensure continued rainfall in agricultural zones in temperate regions, prevent desertification, maintain good air quality and reduce the greenhouse effect, ensure global biodiversity, and improve the living conditions for every person on the planet.

Some nations, among them Costa Rica, welcome the opportunities presented by debt-for-nature swaps. Other nations—Brazil is an example—see the swaps as a form of environmental imperialism in which foreign environmental organizations shape domestic government policy. They argue that restricting large areas of land for parks and nature reserves reduces the amount of land available for economic production and access by poor subsistence farmers.

*Gordon Neal Diem*

### Further Reading

Mahony, Rhona. "Debt-for-Nature Swaps: Who Really Benefits?" 1992. In *Tropical Rainforests: Latin American Nature and Society in Transition*, edited by Susan E. Place. Rev. ed. Wilmington, Del.: Scholarly Resources, 2001.

Meier, Gerald M. *The International Environment of Business: Competition and Governance in the Global Economy.* New York: Oxford University Press, 1998.

Mulder, Monique Borgerhoff, and Peter Coppolillo. "Global Issues, Economics, and Policy." In *Conservation: Linking Ecology, Economics, and Culture.* Princeton, N.J.: Princeton University Press, 2005.

Page, Diana. "Debt-for-Nature Swaps: Experience Gained, Lessons Learned." *International Environmental Affairs* 1 (Fall, 1989): 275-288.

Sachs, Jeffrey. "Making the Brady Plan Work." *Foreign Affairs* 68 (Summer, 1989): 87-104.

## Deforestation

CATEGORIES: Forests and plants; preservation and wilderness issues

DEFINITION: Loss of forestlands through encroachment by agriculture, industrial development, or nonsustainable commercial forestry

SIGNIFICANCE: Deforestation, particularly in tropical regions, has given rise to concerns among environmentalists and scientists, in large part because of the role that tropical forests play in moderating global climate.

Toward the end of the twentieth century, environmentalists became active in decrying the apparent accelerating pace of deforestation because of the loss of wildlife and plant habitats caused by the practice as well as its negative effects on biodiversity. By the 1990's research by mainstream scientists had confirmed that deforestation was indeed occurring on a global scale and that it posed a serious threat to global ecology. Although steps were taken to slow rates of deforestation in the early years of the twenty-first century, the loss of forestlands continued.

Deforestation as a result of expansion of agricultural lands or nonsustainable timber harvesting has occurred in many regions of the world at different periods in history. The Bible, for example, refers to the cedars of Lebanon. Lebanon, like many of the countries bordering the Mediterranean Sea, was thickly forested several thousand years ago. A growing population, overharvesting, and the introduction of grazing animals such as sheep and goats decimated the forests, which never recovered.

Similarly, the forests of Europe and North America have shifted in total area as human populations have changed over the centuries. When the European colonists arrived in the New World, they immediately began clearing the forests. Trees were harvested for building materials and export back to Europe or were simply felled and burned to clear space for farming. In North America, however, as agriculture became increasingly mechanized and farming shifted to the prairies, abandoned farms reverted to woodland. Environmental historians believe, in fact, that a greater percentage of land area in North America is now forested than was covered with trees prior to the arrival of European colonists. A similar phenomenon has taken place in many northern European countries as their populations have become increasingly urbanized.

As the European industrialized nations have gained forestland, however, the less developed countries in Latin America, Asia, and Africa have lost woodlands. While some of this deforestation has been caused by a demand for tropical hardwoods for lumber or pulp, the leading cause of deforestation in the twentieth and twenty-first centuries, as it was several hundred years ago, has been the expansion of agriculture. Growing demand by the industrialized world for agricultural products such as beef has led to millions of hectares of forestland being bulldozed or burned to create pastures for cattle. Researchers in Central America have watched with dismay as large beef-raising operations have expanded into fragile ecosystems in countries such as Costa Rica, Guatemala, and Mexico.

A tragic irony in this expansion of agriculture into tropical rain forests is that the soil underlying the trees is often unsuited for pastureland or raising other crops. Exposed to sunlight, the soil is quickly depleted of nutrients and often hardens. The once-verdant land becomes an arid desert prone to erosion that may never return to forest. As the soil becomes less fertile, thorny weeds begin to choke out the desirable forage plants, and the cattle ranchers move on to clear fresh tracts.

### Slash-and-Burn Agriculture and Logging

Apologists for the beef industry often argue that their ranching practices are simply a form of slash-and-burn agriculture and do no permanent harm. It is true that many of the indigenous peoples in tropical regions have practiced slash-and-burn agriculture for millennia with only minimal impact on the environment. These farmers burn the understory, or low-growing shrubs and trees, to clear small plots of land. Any large trees that survive their fires are cut down with axes and then burned.

Anthropological studies have shown that the small plots these peasant farmers clear can usually be measured in square meters or feet, not in hectares or acres like cattle ranches, and are used for five to ten years. As fertility of the soil declines in one plot, the farmer clears a small plot next to the depleted one. The farmer's family or village gradually rotates through the forest, clearing small plots and using them for a few years, and then shifting to new ground, until they eventually come back to where they began one hundred or more years before. As long as the size of the plots cleared by peasant farmers remains small in proportion to the forest overall, slash-and-burn agriculture does not contribute significantly to deforestation. If the population of farmers grows, however, and more land must be cleared with each succeeding generation, as has been happening in many tropical countries, then even traditional slash-and-burn agriculture can be as ecologically devastating as the more mechanized cattle ranching operations.

Although logging is not the leading cause of deforestation, it remains a significant factor. Tropical forests are rarely clear-cut, as they typically contain hundreds of different species of trees, most of which may have no commercial value. Loggers may select only a few trees for harvesting from each stand. Selective harvesting is a standard practice in sustainable forestry, but just as loggers engaged in the disreputable practice of high-grading across North America in the nineteenth century, so loggers have high-graded in the late twentieth and early twenty-first centuries in Malaysia, Indonesia, and other tropical forests. High-grading is a practice in which loggers cut over a tract to remove the most valuable timber while ignoring the damage being done to the residual stand. The assumption is that, having logged over the tract once, the timber company will not be coming back. This practice stopped in North America not because the timber companies voluntarily recognized the ecological damage they were doing but because they ran out of easily accessible, old-growth timber to cut. Fear of a timber famine caused logging companies to create forest plantations and to undertake the practice of sustainable forestry. While global satellite photos indicate significant deforestation has occurred in tropical areas, enough easily harvested old-growth forest remains in some areas that there is no economic incentive for timber companies to switch to sustainable forestry.

Logging may also contribute to deforestation by making it easier for agriculture to encroach on

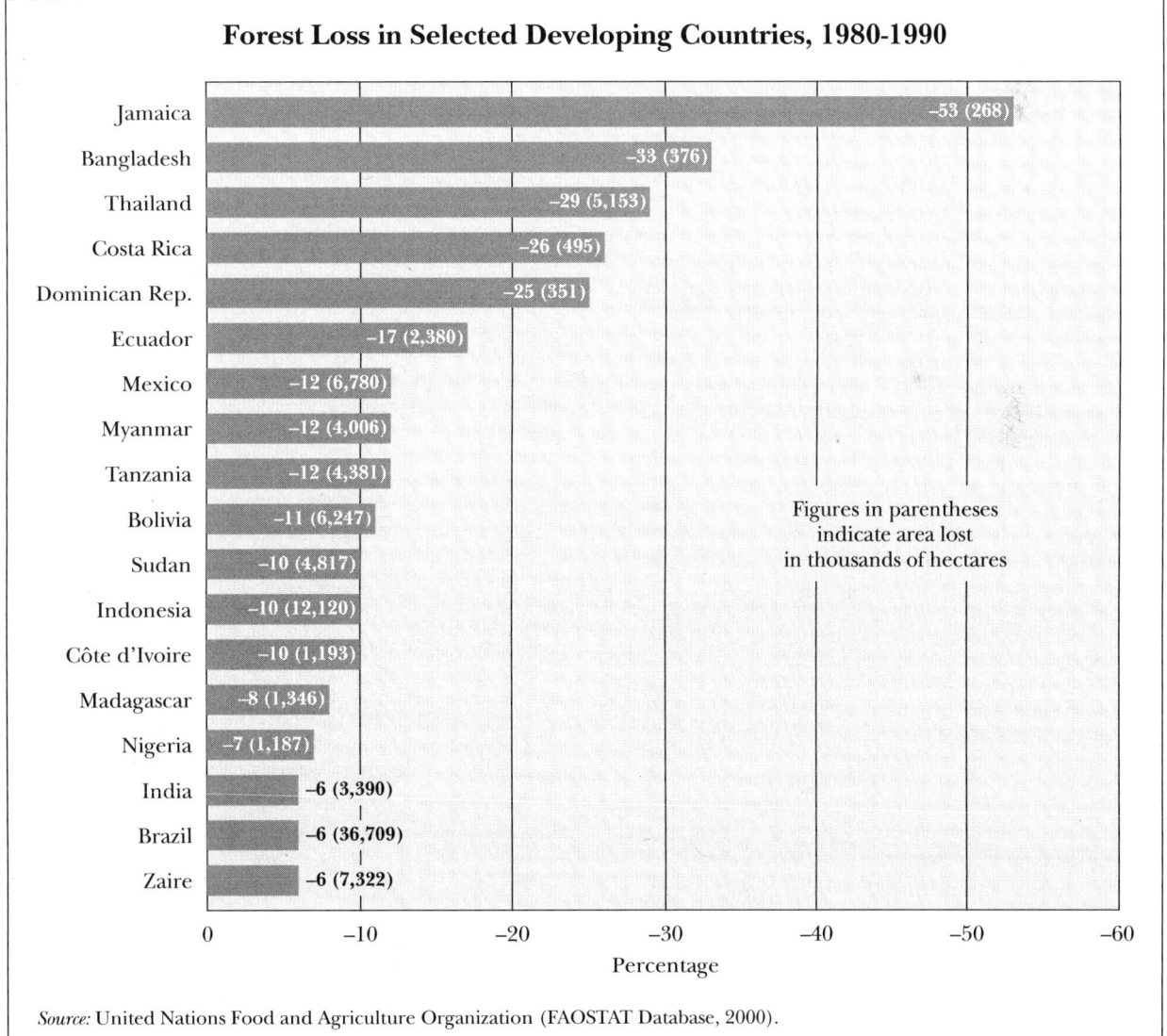

forestlands. Logging companies build roads for their own use while harvesting trees, and farmers and ranchers later use these roads to move into the logged tracts, where they clear whatever trees the loggers have left.

### Environmental Impacts

The extent of the problem of deforestation has long been a subject of debate. The United Nations Food and Agriculture Organization (FAO), which monitors deforestation worldwide, bases its statistics on measurements taken from satellite images. These data indicate that between 2000 and 2005 the net loss of forest area globally was 7.3 million hectares (18 million acres) per year, a reduction from 8.9 million hectares (22 million acres) per year in the decade between 1990 and 2000. In the period 2000-2005, South America and Africa had the greatest losses: South America lost 4.3 million hectares (10.6 million acres) of forest per year, and Africa lost 4 million hectares (9.9 million acres) per year. Environmental activists have been particularly concerned about forest losses in Indonesia and Malaysia, two countries where timber companies have been accused of abusing or exploiting native peoples in addition to engaging in environmentally damaging harvesting methods.

Researchers outside the United Nations have challenged FAO's data, with some scientists claiming the numbers are much too high and others providing convincing evidence that, if anything, FAO's numbers are too low. Few researchers, however, have tried to claim that deforestation on a global scale is not happening. In the 1990's the reforestation of the Northern Hemisphere, while providing an encouraging example that it is possible to reverse deforestation, was not enough to offset the depletion of forestland in tropical areas.

Deforestation affects the environment in a multitude of ways. The most obvious is in a loss of biodiversity. When an ecosystem is radically altered through deforestation, the trees are not the only thing to disappear. Wildlife decreases in number and in variety, and other plants also die. As forest habitat shrinks through deforestation, various plants and animals become vulnerable to extinction. Many biologists believe that numerous animals and plants native to tropical forests will become extinct as the result of deforestation before humans ever have a chance to become aware of their existence.

Other effects of deforestation may be less obvious. Deforestation can lead to increased flooding during rainy seasons. Rainwater that once would have been slowed or absorbed by trees instead runs off denuded hillsides, pushing rivers over their banks and causing devastating floods downstream. The role of forests in regulating water has long been recognized by both engineers and foresters. Flood control was, in fact, one of the motivations behind the creation of the federal forest reserves in the United States during the nineteenth century. More recently, disastrous floods in Bangladesh have been blamed on the logging of tropical hardwoods in the mountains of Nepal and India.

Conversely, trees can also help to mitigate drought. Like all plants, trees release water into the atmosphere through the process of transpiration. As the world's forests shrink, fewer greenhouse gases such as carbon dioxide will be removed from the atmosphere, less oxygen and water will be released into it, and the world will become a hotter, dryer place. Scientists and policy analysts alike are in agreement that deforestation is a major threat to the environment. The question is whether effective policies can be developed to reverse it.

*Nancy Farm Männikkö*

### Further Reading

Chew, Sing C. *World Ecological Degradation: Accumulation, Urbanization, and Deforestation, 3000 B.C.-A.D. 2000*. Walnut Creek, Calif.: AltaMira Press, 2001.

Dean, Warren. *With Broadax and Firebrand: The Destruction of the Brazilian Atlantic Forest*. Berkeley: University of California Press, 1997.

Geist, Helmut J., and Eric F. Lambin. "Proximate Causes and Underlying Driving Forces of Tropical Deforestation." *BioScience* 52, no. 2 (2002): 143-150.

Humphreys, David. *Logjam: Deforestation and the Crisis of Global Governance*. Sterling, Va.: Earthscan, 2006.

Palmer, Charles, and Stefanie Engel, eds. *Avoided Deforestation: Prospects for Mitigating Climate Change*. New York: Routledge, 2009.

Richards, John F., and Richard P. Tucker, eds. *World Deforestation in the Twentieth Century*. Durham, N.C.: Duke University Press, 1990.

Rudel, Thomas K., and Bruce Horowitz. *Tropical Deforestation: Small Farmers and Land Clearing in the Ecuadorian Amazon*. New York: Columbia University Press, 1994.

Sanchez, Ilya B., and Carl L. Alonso, eds. *Deforestation Research Progress*. New York: Nova Science, 2008.

Sponsel, Leslie E., Robert Converse Bailey, and Thomas N. Headland, eds. *Tropical Deforestation: The Human Dimension.* New York: Columbia University Press, 1996.

## Ecotourism

CATEGORY: Preservation and wilderness issues

DEFINITION: Environmentally, socially, and culturally responsible recreational travel intended to preserve ecosystems and improve the well-being of local populations

SIGNIFICANCE: Supporters claim that ecotourism dollars help save endangered wilderness areas that might otherwise be subject to indiscriminate exploitation of natural resources. Detractors note that even the most conscientious ecotourism can contribute to the destruction of fragile ecosystems and cultures, and poor ecotourism practices can wreak even more havoc.

Improvements in travel after World War II, especially the development of jet aircraft, dramatically increased the numbers of tourists in all areas of the globe. With this trend came increased interest in visiting exotic locations to enjoy unspoiled landscapes, view unusual wildlife, and participate in recreational adventures. According to the United Nations World Tourism Organization (UNWTO), by the late twentieth century tourism had become a main income source and the top export category for many developing countries.

The rise in tourism as a leisure activity brought economic benefits such as development and employment opportunities to many pristine areas, but it was sometimes accompanied by negative social, cultural, and environmental impacts. Local communities and lifestyles were sometimes displaced, and ecosystems were altered with the building of hotels, roads, and other amenities for guests. The growing numbers of tourists threatened the very vistas and animals that lured visitors in the first place.

Despite these problems, environmentalists recognize tourism as a means to benefit preservation efforts. While developing countries have the option of exploiting their natural resources to provide revenue, preserving those resources can provide an alternative source of income—tourist dollars—that gives governments an incentive to protect wilderness areas. Coupled with a "no-impact" ethic, ecotourism is seen as a method of saving ecosystems that are quickly disappearing.

Ideally, ecotourism operations employ practices that have minimal negative impacts on the environment and local cultures. Tours focus on natural destinations and rotate the routes they travel and the sites they visit. Participants gain an understanding of their surroundings and how human activity—including their own—affects the ecosystem. Local communities and indigenous populations are involved in managing ecotourism and reap economic benefits from it. The revenues produced by ecotourism are used to help preserve the natural environment.

Ecotourism proponents point to the regions that have successfully used ecotourism to preserve environments and support local communities. Ecotourism in the Ecuadoran rain forest has staved off oil exploration and provides income to native peoples in the area. A former director of a mountain gorilla project in Africa credits ecotourism with the survival of mountain gorillas and their habitats; gorilla ecotourism has also provided significant revenue for local communities. In Costa Rica, the market demand for pristine wilderness has led to the establishment of national parks and protected areas over more than 20 percent of the nation's territory. In Kenya, hundreds of millions of annual tourist dollars provide a powerful incentive to ensure the survival of the country's elephant and rhinoceros populations.

---

### Principles of Ecotourism

*The International Ecotourism Society, the oldest international nonprofit organization devoted to the promotion of ecotourism as a tool for conservation, states that "ecotourism is about uniting conservation, communities, and sustainable travel" and advises that "those who implement and participate in ecotourism activities" should follow these principles:*

- Minimize impact.
- Build environmental and cultural awareness and respect.
- Provide positive experiences for both visitors and hosts.
- Provide direct financial benefits for conservation.
- Provide financial benefits and empowerment for local people.
- Raise sensitivity to host countries' political, environmental, and social climate.

Ecotourism is a growth industry. According to a 2001 report by the Worldwatch Institute, ecotourism grew 20 to 34 percent every year beginning during the 1990's. The UNWTO has reported that in 2004 ecotourism and nature tourism were expanding three times faster than the international tourism industry as a whole.

## Negative Impacts

Ecotourism is not without its drawbacks. Observers in Costa Rica, for example, have noted that although some national parks are large, most visitors want to see specific sites, which leads to overcrowding, trail erosion, and pollution at those sites. Also, scientists have noted changes in the behavioral patterns of local wildlife that appear to be linked to human activity. In Africa, the proximity of ecotourist groups to mountain gorillas puts the great apes at risk from human infectious diseases such as measles, polio, influenza, and tuberculosis.

Growth in ecotourism also promotes development outside protected areas, with attendant environmental degradation. In addition, not all of the people who participate in ecotourism activities have a deep understanding of the no-impact philosophy and a full appreciation of its importance; some of these people contribute to negative impacts through their actions in sensitive areas.

To complicate matters, some purported ecotourism is little more than greenwashed tourism. The burgeoning popularity of ecotourism has led to a proliferation of companies offering purported ecotours that actually fail to employ sustainable practices. In the absence of regulation or even consensus on what constitutes ecotourism, some operators sell their products as ecotours despite the fact that they do not meet the standards of the term as it is usually understood. One Costa Rican tourism project touted as an ecodevelopment included environmentally unfriendly amenities such as a shopping center and a golf course.

Studies indicate that local communities often do not benefit from activities in their surrounding areas touted as ecotourism. In many countries, foreign interests own tourist facilities and recreational sites, thus ensuring that profits flow out of the local area. Environmentally insensitive tourism can displace native populations into marginal lands or drive them from a subsistence lifestyle into poverty-wage service jobs. In Nepal, local families earn little money while serving as porters for tourists. In areas where locals do profit, problems can still arise. Some communities in Costa Rica, for example, have moved from a subsistence to a market economy, a transition that belies the ethic of maintaining the integrity of local cultures.

Critics maintain that the concept of ecotourism is inherently flawed. They argue that ecotourists merely pave the way for mass tourists, people who demand the comforts of home while they visit remote areas. Moreover, the developing nations that offer ecotourist attractions are often the least able to invest the funds necessary to counter the negative impacts of tourism. Only a small percentage of tourist dollars may go toward the management of natural resources.

Opponents of ecotourism assert that it is merely a variant of tourism that will inevitably despoil the very areas it is intended to protect. The deluge of tourists visiting the Galápagos Islands, for example, has overwhelmed the Ecuadoran government's ability to manage them. The annual number of visitors to the islands surpassed the government's target limit of 25,000 people decades ago; by 2005, the number of visitors per year had swelled to more than 121,000. Economic development to accommodate the tourist traffic has caused appreciable damage to the fragile island environment. Environmental advocates recommend that potential ecotourists carefully review the literature of any organization that offers ecotours to be sure that its practices and philosophy are in keeping with the goals of environmental and cultural preservation.

## Emerging Standards

Interest in ecotourism's role in sustainable development and concerns regarding the detrimental effects of ecotourism's mismanagement led to the first World Ecotourism Summit, held in Quebec, Canada. A joint initiative of the UNWTO and the United Nations Environment Programme (UNEP), the summit was held in 2002, designated by the United Nations as the International Year of Ecotourism. The summit laid the groundwork for the Global Sustainable Tourism Criteria (GSTC), introduced in 2008 by the United Nations Foundation, the UNWTO, UNEP, and the Rainforest Alliance. The first international criteria for sustainable tourism practices, these voluntary standards are based on four key elements of sustainable tourism: effective sustainability planning, maximum social and economic benefits for local communities, minimum negative impacts on cultural heritage, and minimum negative impacts on the environment. The criteria are meant not only for ecotourism but also to guide the tourism industry in general toward sustainable practices.

In 2010 the Tourism Sustainability Council, a global membership body, began developing an accreditation program for the world's existing ecotourism certification bodies to bring ecotourism businesses into compliance with universal standards. The program as proposed will use measurable indicators—such as electricity and energy consumption per serviced area, freshwater consumption and waste production per guest per night, and the quality of water discharged from on-site wastewater treatment facilities—to distinguish true ecotourism businesses from greenwashed enterprises.

*Thomas Clarkin*
*Updated by Karen N. Kähler*

FURTHER READING

Fennell, David A. *Ecotourism*. 3d ed. New York: Routledge, 2008.
France, Lesley. *The Earthscan Reader in Sustainable Tourism*. 1997. Reprint. Sterling, Va.: Earthscan, 2002.
Honey, Martha. *Ecotourism and Sustainable Development: Who Owns Paradise?* 2d ed. Washington, D.C.: Island Press, 2008.
McLaren, Deborah. *Rethinking Tourism and Ecotravel*. 2d ed. Bloomfield, Conn.: Kumarian Press, 2003.
Patterson, Carol. *The Business of Ecotourism: The Complete Guide for Nature and Culture-Based Tourism Operators*. 3d ed. Victoria, B.C.: Trafford, 2007.
Schellhorn, Matthias. "Development for Whom? Social Justice and the Business of Ecotourism." *Journal of Sustainable Tourism* 48, no. 1 (2010): 115-136.
Weaver, David B. *Ecotourism*. 2d ed. Milton, Qld.: John Wiley & Sons Australia, 2008.
_____, ed. *The Encyclopedia of Ecotourism*. Oxford, England: CABI, 2001.

## Endangered species and species protection policy

CATEGORIES: Animals and endangered species; forests and plants
DEFINITIONS: Plant and animal species whose numbers are so reduced that the species are in danger of becoming extinct if protection is not provided, and high-level governmental plans of action to support the survival and recovery of endangered and threatened species
SIGNIFICANCE: Natural causes as well as pollution, habitat fragmentation and destruction, and other environmental stresses imposed by human activity can drive species toward extinction. Once a population's size declines past a certain point, various factors will eventually wipe out the population entirely. Implementing protective policies can save a declining species from extinction and, ideally, enable it to recover and thrive.

Extinction of a species does not occur in a vacuum. Causes, typically environmental, are many and often complex. Likewise, because of the many intricate, interconnected relationships existing within ecosystems, the loss of any member may have a ripple effect, eventually having profound negative results. For example, the extinction of a single insect, bird, or bat species may result in the extinction of one or more plant species dependent on the animal species for pollination. If the plant is a critical item in the diet of certain animals, those too may be adversely affected.

Paul R. Erhlich and Anne H. Ehrlich introduce their book *Extinction: The Causes and Consequences of the Disappearance of Species* (1981) by referring to fictitious "rivet poppers"—workers whose job it is to remove rivets from the wings of airplanes. The expectation is that many rivets could be removed without the wings falling off. By analogy, the Ehrlichs consider many world leaders—politicians, bureaucrats, industrialists, engineers, religious leaders, and even some scientists—to be rivet poppers. Through their policies and practices, these leaders espouse programs that will, by design or neglect, result in the loss of endangered species. Ecosystems, by their nature, are somewhat redundant: They are likely to continue to function even after the loss of several species. Ecologists refer to this capacity as "resistance." Ecosystems also possess resilience, or the ability to recover after disturbances, including those in which species are lost. However, just as one would not wish to fly in an airplane from which even a few rivets have been removed, it seems only prudent to take reasonable steps to prevent endangered species from becoming extinct.

Extinction is the conclusion of a long, gradual process typically involving a considerable span of time. When a species undergoes a drastic reduction in the extent of its range, accompanied by a reduction in the number of individuals, it may be designated as a rare species. As this trend continues, the species is likely to be considered threatened prior to being recognized as endangered.

### Factors Contributing to Species Loss

Whether because of their intrinsic nature or environmental conditions, some species are naturally more predisposed to becoming endangered or extinct than others. As one would expect, species with smaller numbers of individuals are more vulnerable than those with more. Each species has a critical population size. Once the numbers fall below that size, the species is especially subject to extinction. Natural populations undergo year-to-year fluctuations in numbers; therefore, a small population will "crash" more readily than a large one.

Several categories of animal and plant species are at high risk of becoming endangered or extinct. Among these are species restricted to special habitats. Most such animal or plant species, by becoming tolerant of an unusual situation, lose their ability to compete in a more general one. One example is island species: If threatened by humans, predators, competing exotic species, or diseases, native island species cannot easily escape. A disproportionate number of animals native to islands have become extinct. Large species with low reproductive rates are also at risk. Large species require more space than do smaller species; therefore, the number of large specimens occupying a given area is lower than the number of smaller ones. Also, most large species, whether whales or trees, are likely to reproduce less often than smaller ones. Even when large species are protected, it is difficult for them to increase their numbers.

Neotropical migratory birds such as warblers, orioles, and tanagers winter in tropical Central or South America or the Caribbean and breed in eastern North America. Their migratory pattern is advantageous in that they can take advantage of the availability of summer food in the north while escaping harsh conditions in winter. Migration, however, is a process that is fraught with danger. As the tropical forests in which they spend the winter are destroyed and the temperate forests in which they breed are fragmented, neotropical migrants may be threatened; thus, they are subject to double jeopardy.

Among the other at-risk species are those at the end of long food chains. Animals such as hawks, owls, and various cat species suffer when any of the links in their food chain are affected. Also, they may be more subject to damage by toxic substances such as environmentally persistent pesticides because of chemical amplification along the food chain. Finally, species of economic value are also in a precarious situation. Many animals have been hunted to extinction; an often-cited example is the passenger pigeon. Plants used medicinally, such as ginseng, have been subjected to overcollecting. Regulatory protections are in place to control the harvesting of American ginseng, which has been dug in eastern North America for centuries.

### Conservation and Management

In order to preserve biodiversity and not lose species that are important to the health and existence of an ecosystem, wildlife conservation and management practices must be put into effect. There are three basic approaches to wildlife conservation and management: the species approach, the ecosystem approach, and the wildlife management approach.

The species approach involves giving endangered species legal protection, protecting and managing their habitats, propagating species in captivity, and reintroducing species into safe habitats. In 1903 President Theodore Roosevelt established the first wildlife refuge in the United States. The refuge, located on Pelican Island on the east coast of Florida, was developed to protect the brown pelican, which was in decline. (Only in 2009 was the species officially declared to be out of danger.) Since then the National Wildlife Refuge System has grown to more than 550 refuges and other units. Habitats in the United States are also protected through the national park and forest systems and the National Wilderness Preservation System. In addition to the government, private conservation organizations such as the National Audubon Society, the Sierra Club, and the Nature Conservancy have been of tremendous value in acquiring and protecting sensitive landscapes.

According to statistics reported by the International Union for Conservation of Nature and Natural Resources (IUCN), as of 2000 about thirty thousand protected areas had been established around the world. These areas, which include strict nature reserves and wilderness areas, national parks, natural monuments, habitat and species management areas, protected landscapes and seascapes, and managed resource protected areas, represent more than 13,250,000 square kilometers (5,115,854 square miles) of the planet's land surface.

Other forms of the species approach to saving diversity include gene banks, botanical gardens, and zoos. The seeds of many endangered plant species are preserved in climatically controlled environments. The organization Botanic Gardens Conservation International estimates that more than eighty thousand plant species are in cultivation in the world's botanic gardens. Many botanical gardens, such as Kew Gardens in England, are repositories for plant species that are endangered or have even ceased to exist in the wild. Some of these plants are reintroduced into native habitats after being cultivated for decades in these gardens or in seed banks.

Egg pulling and captive breeding are two methods that zoos and animal research centers use for preserving endangered animal species. Egg pulling involves collecting eggs from endangered species in the wild and hatching the eggs in zoos or research centers, as was done with California condors beginning in 1983. Endangered species still in the wild are sometimes captured and put into research centers to breed in a controlled environment. When the captive populations become large enough, some of the individuals are reintroduced into protected habitats. The Arabian oryx, a large antelope species that originated in the Middle East, was hunted to extinction in the wild; however, the species survived thanks to captive breeding programs that began in San Diego, Los Angeles, and Phoenix zoos. The oryx has since been reintroduced into its native habitats, although the captive population outnumbers the population in the wild.

The second approach to saving biodiversity is the ecosystem approach, which emphasizes preserving balanced populations of species within their native habitats. It involves establishing legally protected wilderness areas and wildlife reserves. An important part of making sure that the habitat is safe is to eliminate all alien species from the area. The Minnesota Zoo has formed a partnership with other organizations to help protect certain animals in their native habitats, notably the desert black rhino and Hartmann's mountain zebra. Instead of moving these animals to Minnesota, the zoo supports conservation efforts to study and protect these animals in their native habitats in Namibia, Africa.

The third approach to preserving biodiversity is the wildlife management approach. When it is decided which species or group of species will be managed in a given area, a management plan is put into effect. Steps in the plan include investigating and determining the kinds of cover, food, water, and space the targeted species requires. Action is then taken to grow the plants that provide the needed cover and food for the species.

### Hunting and International Cooperation

Legal and illegal commercial hunting has led to the extinction or near extinction of many animal species. Despite policies to regulate hunting, poaching remains a lucrative business, particularly in underdeveloped countries. Some threatened and endangered species are killed for their hides, horns, or other ornamental or medicinal parts, while others are captured and smuggled alive, as there is a market for exotic pets and decorative plants.

### Endangered and Threatened Species, 2008

| | Mammals | Birds | Reptiles | Amphibians | Fishes | Snails | Clams | Crustaceans | Insects | Arachnids | Plants |
|---|---|---|---|---|---|---|---|---|---|---|---|
| **Total listings** | 357 | 275 | 119 | 32 | 151 | 76 | 72 | 22 | 61 | 12 | 747 |
| *Endangered species* | 325 | 254 | 79 | 21 | 85 | 65 | 64 | 19 | 51 | 12 | 599 |
| United States | 69 | 75 | 13 | 13 | 74 | 64 | 62 | 19 | 47 | 12 | 598 |
| Other countries | 256 | 179 | 66 | 8 | 11 | 1 | 2 | — | 4 | — | 1 |
| *Threatened species* | 32 | 21 | 40 | 11 | 66 | 11 | 8 | 3 | 10 | — | 148 |
| United States | 12 | 15 | 24 | 10 | 65 | 11 | 8 | 3 | 10 | — | 146 |
| Other countries | 20 | 6 | 16 | 1 | 1 | — | — | — | — | — | 2 |

Source: Data from U.S. Department of Commerce, *Statistical Abstract of the United States, 2009*, 2009.
Note: Numbers reflect species listed by U.S. government as "threatened" or "endangered"; actual worldwide totals of species that could be considered threatened or endangered are unknown but are believed to be higher.

Species, primarily game species, are managed through the establishment of laws that regulate hunting and hunting quotas. Hunters are required to have licenses and to use only certain types of hunting equipment and are permitted to hunt only during certain months of the year. Limits are set on the size, number, and sex of animals that can be hunted in a given game refuge.

Management plans and international treaties have been developed to protect migrating game species, such as waterfowl. In North America, waterfowl such as ducks, geese, and swans nest in Canada during the summer and migrate to the United States and Central America in the fall and winter. The United States, Canada, and Mexico have signed agreements to protect these waterfowl from overhunting and habitat destruction.

Some waterfowl refuges in the United States include human-built nesting sites, ponds, and nesting islands. The U.S. National Wildlife Refuge System includes thirty-eight wetland management districts administering more than twenty-six thousand waterfowl production areas, which contribute to the protection of migratory birds. In 1986, amid concerns regarding record lows in waterfowl populations, the United States and Canada entered into an agreement (later joined by Mexico) to attempt to restore the continental waterfowl population and associated habitats. By the end of 2009, $4.5 billion had been invested to protect, restore, and enhance 6.3 million hectares (15.7 million acres) of waterfowl habitat in North America.

In July, 1975, a wide-reaching treaty to protect endangered species, the Convention on International Trade in Endangered Species of Wild Fauna and Flora (CITES) went into force. This agreement, which by 2010 had been agreed to by 175 national governments, extends varying protections to more than thirty thousand plant and animal species. Under CITES, some endangered and threatened species cannot be commercially traded, either alive or as products. Others can be traded, but only by persons who obtain the proper export licenses. One of the best-known results of the CITES agreement is the 1989 ban (subsequently weakened) on the international trade in ivory. The ban was enacted to halt the decline of the African elephant, which had dwindled in population from 2.5 million animals in 1950 to approximately 350,000 at the treaty's inception.

In 1980 the World Wildlife Fund (now the World Wide Fund for Nature), the United Nations Environment Programme, and IUCN developed a world conservation strategy. The plan, which was expanded in 1991, seeks to preserve biological diversity, combine wildlife conservation and sustainable development, encourage rehabilitation of degraded ecosystems, and monitor sustainability of ecosystems. Fifty countries have established national conservation programs in response to this plan.

U.S. Laws

Important U.S. laws that control imports and exports of endangered wildlife and wildlife products began with the Lacey Act of 1900, passed in response to an egret population decline resulting from the commercial value of their feathers as decoration. The Lacey Act prohibited transporting live or dead wild animals or their parts across state borders without a federal permit. Later came the Endangered Species Preservation Act of 1966, then the Endangered Species Act (ESA) of 1973, which has been amended several times. The ESA was unique in that, where previous wildlife regulations had focused primarily on game animals, the ESA program focused on identification of all endangered species and populations in order to save biodiversity, regardless of the species' usefulness to humans.

The act classifies endangered species as those that are in immediate danger of extinction and threatened species as those that are likely to become endangered in a given habitat in the future. Some species are classified as being locally threatened even though they can be found in fairly large numbers in some parts of their former habitats. The U.S. Fish and Wildlife Service is required to prepare a recovery plan for each species that the ESA lists as officially endangered. In 2010 the ESA listings for endangered and threatened U.S. species included 578 animals and 793 plants.

The ESA provides that a listed species cannot be harassed, harmed, pursued, hunted, shot, trapped, killed, captured, or collected, either on purpose or by accident. It further prohibits importing or exporting endangered species, as well as possessing, selling, transporting, or offering to sell any endangered species. Violators face fines and imprisonment. In 1995 the U.S. Supreme Court ruled to extend further protection for endangered species by ruling that habitat essential for species survival must be protected, whether on public or private land.

Critics view the Endangered Species Act as a major stumbling block to economic progress. A classic example is the delay that occurred in the late 1970's in the construction of Tellico Dam in the mountains of eastern Tennessee because of the presence of the snail darter, a small endangered fish. The dam was ultimately completed, and other populations of the fish were unexpectedly found elsewhere, resulting in the snail darter's being removed from the endangered list. Some people who criticize the ESA assert that business and public interests should take precedence over the protection of wildlife, especially plants and animals of no obvious value to human beings.

Some of the ESA's recovery plans have proved successful. Bald eagles, which numbered only 800 in 1970, were able to rebound to a population of almost 9,800 by 2006, largely because of the U.S. ban on the pesticide dichloro-diphenyl-trichloroethane (DDT). The American alligator was listed as an endangered species in 1967 after its population declined because of habitat destruction and high demand for alligator meat and products made from alligator hides. Because of ESA protection, the alligator was reestablished in its southern range; it was removed from the endangered list in 1987. Most of the ESA success stories have involved such "charismatic megafauna." Less glamorous endangered species, such as fungi, wildflowers, liverworts, mosses, and insects, receive less attention even though their roles in ecosystems may be more important. In spite of ESA protection, a number of species remain critically endangered, largely because their standing was so precarious by the time they were listed as protected. ESA listed status has not saved some species from going extinct.

## OTHER PROTECTIVE MEASURES

The loss of aquatic species has generally attracted less attention than the extinction of land species. With the realization of the importance of healthy freshwater and marine species, however, governments have begun establishing marine preserves. Fishing, construction, tourism, pollution, and other human disturbances are closely regulated and restricted in these areas. The National Marine Sanctuary Program, which was developed in 1972 in the United States, has established fourteen marine protected areas. The most extensive of these, the Papahānaumokuākea Marine National Monument in the northwest Hawaiian Islands, is the single largest conservation area in the United States and the world's largest fully protected marine area. It includes both marine and terrestrial habitats.

International measures to protect species from destruction or exploitation include the 1946 International Convention for the Regulation of Whaling. Overwhaling worldwide caused a huge decline in whales, from an estimated 4.4 million in 1900 to approximately 1 million by the end of the twentieth century. Overharvesting of whales affected almost every whale species of commercial value. In 1946 the International Whaling Commission (IWC) was established to set annual whaling quotas to prevent commercial overharvesting and the extinction of whales. However, many whaling countries ignored the suggested quotas. In 1971 the United States stopped all commercial whaling and banned imports of all whale products. In 1974 the IWC began to regulate whaling according to the principle of maximum sustainable yield. When a given species of whale—such as the right whale, bowhead whale, or blue whale—fell below the optimal population for such a yield, the IWC issued a ban on hunting that species.

Other international agreements to protect endangered species include the Convention on the Conservation of Migratory Species of Wild Animals (Bonn Convention), which entered into force in 1983, and the Convention on Biological Diversity (CBD), which entered into force in 1993. The Bonn Convention is the only international treaty that focuses on the conservation of terrestrial, marine, and avian migratory species, their habitats, and their migration routes. The CBD, which opened for signature at the 1992 Earth Summit in Rio de Janeiro, Brazil, concerns the significance of biodiversity for future generations, the sovereignty of each nation over its resources, and each nation's need and right to conserve and protect its own biodiversity. The treaty details a plan that directs industrialized countries to help fund projects for the protection of biodiversity within developing countries; it also stresses the right of national governments to decide who may have access to their resources. The treaty provides for a sharing of technologies, particularly biotechnologies that have been developed from plants originating in developing countries, thereby giving developing countries substantial benefits from any technologies that are developed based on theses countries' genetic resources. Agenda 21, a comprehensive international plan of action adopted at the Earth Summit, addresses the need for nations to "promote the rehabilitation and restoration of damaged ecosystems and the recovery of threatened and endangered species."

*Thomas E. Hemmerly, Toby Stewart, and Dion Stewart*
*Updated by Karen N. Kähler*

FURTHER READING

Bräutigam, Amie, and Martin Jenkins. *The Red Book: The Extinction Crisis Face to Face.* Gland, Switzerland: IUCN, 2001.

Chiras, Daniel D. "Preserving Biological Diversity." In *Environmental Science.* 8th ed. Sudbury, Mass.: Jones and Bartlett, 2010.

Groom, Martha J., Gary K. Meffe, and Carl Ronald Carroll. *Principles of Conservation Biology.* 3d ed. Sunderland, Mass.: Sinauer, 2006.

Mackay, Richard. *The Atlas of Endangered Species.* Rev. 3d ed. Sterling, Va.: Earthscan, 2009.

McNeely, Jeffrey A., et al. *Conserving the World's Biological Diversity.* Washington, D.C.: Island Press, 1990.

Vié, Jean-Christophe, Craig Hilton-Taylor, and Simon N. Stuart, eds. *Wildlife in a Changing World: An Analysis of the 2008 IUCN Red List of Threatened Species.* Gland, Switzerland: IUCN, 2009.

Wagner, Viqi, ed. *Endangered Species: Opposing Viewpoints.* Detroit: Greenhaven Press, 2008.

# Environmental impact assessments and statements

CATEGORY: Land and land use

DEFINITION: Evaluations of the likely impacts of proposed or existing human activities on given environments, and the reports of the findings in documents for public review

SIGNIFICANCE: Implementation of legal requirements that environmental impact assessments be performed and statements of the results be made public have resulted in changes in construction, resource extraction, and land-use planning that take into account the larger environment.

Various kinds of human activities may trigger environmental impact assessments (EIAs); these include construction, resource extraction, and land-management policy implementation. An entity planning a development can choose to do an EIA, but formal EIAs are mandated responses to specific legislation. Under some legislation, an initial scoping process is done to determine whether a more lengthy process, the EIA, is required. The results of a legally mandated or regulatory EIA are documented in an environmental impact statement (EIS).

The mandating of EIAs has become increasingly common in the United States and in other nations because of intensified public pressure on legislators and policy makers to improve resource management and conservation. The U.S. National Environmental Policy Act of 1969 (NEPA) helped usher in the era of environmental assessment for government decision making. By 1998 more than one hundred countries had established EIA processes. Ideally, administration of an EIA program promotes government and corporate accountability for environmental alterations and institutionalizes systematic, science-based policy analysis.

The environmental movement, spurred by Rachel Carson's book *Silent Spring* (1962), influenced U.S. legislators to reconsider the lack of a national policy for the environment. NEPA, signed into law on January 1, 1970, heralded a new role for citizens to participate in reviews of government decisions. The EIA process produces a draft environmental impact statement (DEIS) and a final environmental impact statement (FEIS) for public comment. However, once the FEIS has been accepted, NEPA has been satisfied, but this in itself does not constitute approval or denial of a proposed project. By focusing on the process rather than the end result (as most environmental permit programs do), the EIA reflects a compromise between environmental and political interests. The goal is to ensure that a suitable EIS or "finding of no significant impact" (FONSI) is prepared. A well-planned project should be able to withstand the public scrutiny. Other laws and permitting processes may be required before a proponent can actually go ahead with a proposed development or action, but these processes can build upon or use the data gathered in the EIA.

In the United States, NEPA marked a change for federal agencies because it added environmental accountability to every agency's mission, along with a specific method to carry out environmental reviews. NEPA not only provided a common thread among agencies but also comprehensively linked various categories or media in which environmental impacts occur. In the EIA process, impacts ranging from archaeological resource depletion to air pollution are examined. Social impacts resulting from proposed actions are as much a part of the EIA as are issues related to water resources, noise, solid waste disposal, and other common types of environmental impacts. Aesthetics has also proved an important if initially nebulous category of impact, although a significant body of literature has arisen to treat the need to quantify impacts normally considered subjective. Socioeconomic values of environmental resources (for life support, amenities, and raw materials) are also used in evaluating trade-offs among alternatives. Formal EIAs in most countries generally include consideration of similar wide arrays of environmental impacts.

The broad categories of impacts are intended to reflect the interconnectedness of environmental settings and allow the interplay of social, economic, political, and environmental issues. The breadth also extends to the type of projects that require environmental assessment. In the United States, projects that are subject to the EIA process include any undertaken by a federal agency; any involving a federal license, permit, or funding; and any taking place on federal property. Approximately two dozen U.S. states have their own equivalent assessment requirements. Nations that use the EIA process are more readily able to participate in global trade, qualify for funding, and meet the increasing international demand for and appreciation of environmental quality.

An EIA assesses more than the proposed action; it also looks at the impacts of legitimate alternatives to the action, as well as the impacts of not doing the project. The EIA process includes comparing the costs and benefits of the alternatives and the various impacts. The EIS documents the impacts, costs, and benefits for the project and its alternatives. Under NEPA-type legislation, a DEIS is circulated, and public comments are solicited either in writing or at hearings. The FEIS is issued after consideration of public and other agency input. The courts provide a forum for class-action suits and other assessment-related disputes. The majority of challenges have been based on allegations of either failure to prepare an EIS or failure to consider fully the proper alternatives. From a public policy perspective, as well as from the perspective of peer-reviewed science, it is the public nature of the EIS that determines the success of the EIA process.

The EIA ideally is part of the planning process rather than an afterthought for a project that has already commenced; however, the EIA must be conducted late enough in the planning stages that a sufficient description of the project exists to allow assessment of the impacts. One response to this problem is the strategic environmental assessment (SEA). The forecasting of impacts, especially for SEAs, cannot simply be done through direct observation. In conducting SEAs, teams of professionals use physical models, mathematical models, qualitative models, checklists, and expert opinions.

Starting with a project description, an EIS proceeds to identification of associated or expected direct and indirect impacts. Next, the existing environmental conditions are described. Then relevant laws and regulations are examined for standards and applicability. Specific environmental impacts are predicted, and their significance is evaluated. The final step is description of the incorporation of the EIA's results into the project to reduce or mitigate the impacts; this includes information on monitoring, reporting, and responding to postconstruction impacts. EIA and EIS notifications appear in legal sections of major newspapers, on government and corporate Internet sites, and at agency offices. EIS's are generally available from the preparers upon request or can be viewed by concerned citizens at various public locations.

*Robert M. Sanford and Hubert B. Stroud*

FURTHER READING

Cantor, Larry W. *Environmental Impact Assessment*. New York: McGraw-Hill, 1996.

Garb, Yaakov, Miriam Manon, and Deike Peters. "Environmental Impact Assessment: Between Bureaucratic Process and Social Learning." In *Handbook of Public Policy Analysis: Theory, Politics, and Methods*, edited by Frank Fischer, Gerald J. Miller, and Mara S. Sidney. Boca Raton, Fla.: CRC Press, 2007.

Glasson, John, Riki Therivel, and Andrew Chadwick. *Introduction to Environmental Impact Assessment*. 3d ed. New York: Routledge, 2005.

Rogers, Peter P., Kazi F. Jalal, and John A. Boyd. "Environmental Assessment." In *An Introduction to Sustainable Development*. Sterling, Va.: Earthscan, 2008.

# Environmental law, international

CATEGORIES: Treaties, laws, and court cases; preservation and wilderness issues; resources and resource management; human health and the environment

# Environmental law, international

DEFINITION: Internationally accepted rules governing the conduct of sovereign nations in their relationships with one another for the international regulation of environmental issues aimed at environmental protection and sustainability

SIGNIFICANCE: International environmental laws regulate the interactions of human beings with the rest of the biophysical or natural environment, with the aim of reducing the negative impacts of human activity on the natural environment.

International environmental laws promote responses to serious threats to the global environment and address some of the most challenging global issues, especially in areas of trade and environment, habitat protection, climate change, oceans and fisheries, water scarcity, endangered species, and access to biological resources. Because pollution does not respect international or political boundaries, environmental law is an important aspect of international relations.

Encompassing agreements among nations, international environmental laws fall into two broad categories: pollution control and remediation, and resource conservation and management. The boundaries of international environmental law are difficult to determine, and therefore it cannot usually be restricted to one specific definition. Rather, it is interdisciplinary and also overlaps and intersects with numerous other areas of research, including ecology, human rights, economics, political science, and navigation.

International environmental law also engages with global efforts to develop commonly accepted environmental and social standards and to govern major investments in developing countries. These efforts involve lawyers and practitioners in the field of environmental law who help to design and enforce global environmental standards.

While the international bodies that proposed, argued, agreed upon, and ultimately adopted existing international agreements vary according to each agreement, certain conferences and commissions have been particularly important, including the United Nations Conference on the Human Environment (1972), the World Commission on Environment and Development (1983), the United Nations Conference on Environment and Development (1992), the World Summit on Sustainable Development (2002), and the United Nations Climate Change Conference, which produced the Copenhagen Accord (2009).

## HISTORY AND BACKGROUND

Environmental law may be said to have begun in 1306, when England's King Edward I banned the burning of coal in London because of the polluting smoke that it produced. Until the late 1960's, most international agreements that aimed at protecting the environment were only narrowly defined. In 1972, the Stockholm Declaration of the United Nations Conference on the Human Environment paved the way for international agreements that began to promote legislation for the preservation of the environment.

The twentieth century movement known as environmentalism brought about many environmental protection laws at the local, state, national, and international levels. These laws create liabilities for businesses that pollute air, water, or soil or improperly dispose of waste. Legally binding international agreements deal with many types of environmental issues, stretching from terrestrial, freshwater, marine, and atmospheric pollution to the protection of biodiversity and wilderness. Numerous international principles and rules have emerged since the late twentieth century to challenge many of the established rules and principles in the international legal field. Several multilateral environmental agreements have been adopted, and international environmental rules regulate almost every environmental issue imaginable, including marine pollution, hazardous activities, atmospheric pollution, waste management, and access to environmental information.

International environmental law in the twenty-first century has focused in particular on such global environmental problems as the depletion of the ozone layer, transboundary movements of hazardous wastes, the conservation of biological diversity, and the international response to climate change. Principles of ecology, conservation, stewardship, responsibility, and sustainability have all influenced modern international environmental law. Numerous international conventions and other agreements set environmental law concerning many issues, including climate change and global warming (the United Nations Framework Convention on Climate Change, the Kyoto Protocol, and the Copenhagen Accord), sustainable development (the Rio Declaration on Environment and Development), biodiversity (the U.N. Convention on Biological Diversity), transfrontier pollution (the U.N. Convention on Long-Range Transboundary Pollution), marine pollution (the U.N. Convention on the

Prevention of Marine Pollution by Dumping of Wastes and Other Matter), endangered species (the U.N. Convention on International Trade in Endangered Species, or CITES), hazardous materials and activities (the Basel Convention on the Control of Transboundary Movements of Hazardous Wastes and Their Disposal), cultural preservation (the Convention Concerning the Protection of the World Cultural and Natural Heritage), desertification (the U.N. Convention to Combat Desertification), and uses of the seas (the U.N. Convention on the Law of the Sea, or UNCLOS).

Global climate change resulting from greenhouse gas emissions has become a particularly important issue in international environmental law. The United Nations Climate Change Conference, held in December, 2009, produced the Copenhagen Accord, which was drafted by the United States, China, India, Brazil, and South Africa. The resulting arguments and unanswered questions caused the U.S. government to agree only tentatively to the proposal, which was not unanimously and officially adopted by all participating countries. The participating countries drafted a document that emphasizes the importance of climate change as one of the most challenging environmental issues of the twenty-first century. Although not yet legally binding, the accord asserts that decisive actions should be taken to limit global temperature increases to less than 2 degrees Celsius (3.6 degrees Fahrenheit). No consensus was reached regarding the reduction of emissions of carbon dioxide—acknowledged by many scientists to be a chief factor in global warming (carbon dioxide is the most important of the "greenhouse gases," gases that contribute to the atmosphere's trapping of radiant heat). Although the contents of the Copenhagen Accord have been opposed by many countries and nongovernmental organizations, and many critics of the accord have asserted that it is only a weak version of what is needed to achieve meaningful action on climate change, by January, 2010, at least 138 countries had signed the agreement.

## International Environmental Law Organizations

Around the world, intergovernmental and nongovernmental organizations seek to promote the adoption and enforcement of internationally accepted environmental laws. Among these organizations are the following:

- Intergovernmental Panel on Climate Change (IPCC): This panel was established by the World Meteorological Organization and the United Nations Environment Programme. The goal of the IPCC is to engage in total and objective assessment of all scientific, technical, and socioeconomic information relevant to an understanding of the scientific basis of the risks posed by climate change induced by humans, the potential impacts of such risks, options for adaptation to climate change, and ways to reduce the effects of climate change.
- Commission for Environmental Cooperation (CEC): This commission, which was created by the North American Agreement on Environmental Cooperation, a side treaty of North American Free Trade Agreement, addresses regional environmental concerns, helps prevent potential trade and environmental conflicts, and promotes the effective enforcement of environmental laws. Members of the CEC are the United States, Canada, and Mexico.
- United Nations Environment Programme (UNEP): UNEP was established as the environmental conscience of the United Nations system. It provides an integrative and interactive mechanism through which a large number of separate efforts by intergovernmental, nongovernmental, national, and regional bodies in the service of the environment are reinforced and interrelated.
- International Maritime Organization (IMO): This organization, which handles marine pollution regulations, is the specialized agency of the United Nations responsible for improving maritime safety and preventing pollution from ships.

Other intergovernmental organizations with interests in international environmental law include the United Nations Commission on Sustainable Development, the United Nations Educational, Scientific, and Cultural Organization's World Heritage Center, the European Commission's Environmental Directorate General, the European Environment Agency, the World Conservation Union, the International Union for Conservation of Nature, the World Bank, and the World Meteorological Organization.

The Center for International Environmental Law (CIEL) is an example of a nongovermental organization that works to support the creation and enforcement of environmental laws around the world. This public interest nonprofit law firm, which has offices in the United States and Switzerland, is committed to strengthening existing international environmental laws and using international law and institutions to protect the environment, promote human health, and ensure a just and sustainable society. Other examples of nongovernmental organizations that support international environmental law include the International Environmental Law Committee of the American Bar Association (which considers, informs, and engages its members on public international environmental law, such as global, multilateral, regional and bilateral agreements on the environment), Earthjustice (formerly the Sierra Club Legal Defense Fund), the EnviroLink Network, the Environmental Law Alliance Worldwide, Greenpeace International, the Natural Resources Defense Council, Friends of the Earth International, Resources for the Future, the World Resources Institute, and the World Wide Fund for Nature.

### Controversies

The definition of "international environmental law" can sometimes generate controversy. Given the broad scope of environmental law, no fully definitive list of environmental laws is possible; the field is made up of a complex of interlocking global, international, national, state, and local treaties, conventions, statutes, regulations, and policies that seek to protect the environment and natural resources affected or endangered by human activities. Together, these instruments attempt to tackle some of the world's most challenging and controversial environment-related issues, including climate change, pollution, resource conservation, trade policies, management of fisheries, and biodiversity preservation.

Controversy arises from the nexus of these goals and economic priorities: From an economic perspective, environmental law may be understood as concerned with the prevention of present and future environmental problems and the preservation of common resources in order to avoid their exhaustion. The limitations and expenses that international environmental laws may impose on commerce are often discussed in the context of the often immeasurable financial benefits of environmental protection. Where policies to promote long-term but difficult-to-measure environmental priorities conflict with policies that promote short-term economic gains (particularly during economic downturns such as the global recession that began in 2008), persons who are suffering economically often find economic goals more compelling and easier to understand than environmental goals.

The concept of environmental justice is also a source of controversy. The development, implementation, and enforcement of environmental laws, regulations, and policies can have disproportionate impacts on particular socioeconomic, ethnic, or national groups of people. Therefore, one of the goals of environmental law must be to clearly define and equitably administer environmental justice for all nations and peoples, by providing each with the same degree of protection from environmental and health hazards and equal access to the decision-making process that seeks to provide healthy environmental living conditions.

Among scientists and politicians, debates are ongoing regarding the existence and causes of climate change—particularly whether it is mainly the result of human activities. If long-term global warming is truly occurring, it will have significant future impacts on agriculture, sea levels, the spread of disease, the deterioration of polar ice (the world's "air conditioner"), species (including human) migration, and biological diversity. If, in addition, global warming can be determined to be primarily the result of human activities (as the IPCC maintains), then the argument can be made that a reduction, or at least an arrest, of human activities contributing to climate change must be implemented. The challenge then becomes how to regulate human-induced global warming with minimal interference in the global economy—and how to do so with minimal disproportionate impacts on particular populations.

That challenge requires creative and innovative law and policy making at all levels of society. Governments around the world must continue to discuss how best to combat climate change and adapt to its effects. Appropriate international environmental laws must be formulated and enforced if goals for the mitigation of negative climate change impacts are to be met. There is a need to gain not only the theoretical but also the proactive support of developing countries for the 2009 Copenhagen Accord, by allowing underprivileged nations their total inputs and helping them to gain an understanding of how to comply with the agreement's complex or misunderstood provisions.

*Samuel V. A. Kisseadoo*

### Further Reading

Birnie, Patricia, Alan Boyle, and Catherine Redgwell. *International Law and the Environment.* 3d ed. New York: Oxford University Press, 2009.

Bodansky, Daniel. *The Art and Craft of International Environmental Law.* Cambridge, Mass.: Harvard University Press, 2010.

Guruswamy, Lakshman D., with Kevin L. Doran. *International Environmental Law in a Nutshell.* 3d ed. St. Paul, Minn.: Thomson/West, 2007.

Hunter, David, James Salzman, and Durwood Zaelke. *International Environmental Law and Policy.* 3d ed. New York: Foundation Press, 2006.

Louka, Elli. *International Environmental Law: Fairness, Effectiveness, and World Order.* New York: Cambridge University Press, 2006.

Sands, Philippe. *Principles of International Environmental Law.* 2d ed. New York: Cambridge University Press, 2003.

Weiss, Edith Brown, et al. *International Environmental Law and Policy.* 2d ed. Frederick, Md.: Aspen Law & Business, 2006.

## European Diploma of Protected Areas

CATEGORY: Preservation and wilderness issues
IDENTIFICATION: Award given to selected European regions that satisfy certain scientific, aesthetic, or cultural criteria
SIGNIFICANCE: The European Diploma of Protected Areas encourages the protection and preservation of selected regions of Europe and recognizes excellence in environmental success.

The Council of Europe established the European Diploma of Protected Areas in 1965. The diploma is awarded based on a number of criteria, including the following: The area must be of importance for the conservation of Europe's biological diversity (whether because it is home to a large number of species or is the habitat of endangered or threatened species) or for the preservation of the continent's landscape diversity, or it must be a site of remarkable natural or geographic phenomena (such as spectacular geological sites, noteworthy paleontological sites, or other historical areas).

Areas that are awarded the diploma must be protected by the laws of the countries in which they are found, must be clearly designated and maintained by specific plans, and must have sufficient staff and resources for their protection. Managers of the areas supervise the protected zones to ensure their maintenance and meet regularly together to discuss common concerns. By 2010 the Directorate of Culture of the Council of Europe had awarded the diploma to almost seventy areas in more than two dozen countries, including Austria, Belarus, Belgium, the Czech Republic, Finland, France, Germany, Greece, Hungary, Italy, Luxembourg, the Netherlands, Romania, Russia, Spain, Sweden, Switzerland, Ukraine, and the United Kingdom.

The awarded areas include the Karlstejn National Nature Reserve in the Czech Republic. The reserve is an important site both biologically and historically, not only for the country but also for the continent as a whole. It contains a number of unique species, especially insects and mollusks, some of which are endangered and others of which are at the northern limit of their habitat. The area has been inhabited for some three thousand years and is the site of the famous Karlstejn castle, built in the fourteenth century. Other protected areas are the Fair Isle National Scenic Area in Scotland, an almost treeless island between Shetland and Orkney; the Minsmere Nature Reserve in eastern England; Cretan White Mountains National Park in Greece; and Boschplaat Nature Reserve in the Netherlands.

One of Germany's awarded protected areas is the notorious Berchtesgaden National Park, where Adolf Hitler had his palatial mountain retreat. The directorate awarded the diploma to this region because of "the exceptional quality of its landscapes, the richness of its flora and fauna and the diversity of its natural sites."

Another area, the only one shared by two countries, is the Germano-Luxembourg Nature Park, which straddles the Our and Sauer rivers. The area is undisturbed by industrialization and contains many cultural and natural sites of interest. Its landscape of high plateaus and ravines in the north and hills and streams in the south captures the features of the Ardennes and Lorraine areas.

*Frederick B. Chary*

FURTHER READING
Hunkeler, Pierre. *European Diploma for Protected Areas.* Strasbourg, France: Council of Europe, 2000.
Keulartz, Josef, and Gilbert Leistra, eds. *Legitimacy in European Nature Conservation Policy: Case Studies in Multilevel Governance.* New York: Springer, 2007.

# Fish and Wildlife Service, U.S.

CATEGORIES: Organizations and agencies; animals and endangered species
IDENTIFICATION: Federal government bureau responsible for protecting fish and animal habitats
DATE: Established in 1940
SIGNIFICANCE: The U.S. Fish and Wildlife Service traces its roots to some of the world's oldest natural resource conservation programs. The agency is responsible for conserving, protecting, and enhancing fish, wildlife, and plant species and their habitats, and its environmental stewardship includes administration of the Endangered Species Act.

During the late nineteenth century, resource use in the United States developed at such a rapid pace that it led to abuse of the environment. Among the resources most depleted were the national fisheries. Major declines in fish populations and commercial fish catches led the U.S. Congress to establish the Commission on Fish and Fisheries in 1871. Its initial responsibilities were to study and reverse the decline in food fishes, restore streams, improve fisheries, and stock lakes and streams.

Concern for wildlife in relation to agriculture also led to the establishment in 1885 of the Division of Economic Ornithology and Mammalogy within the U.S. Department of Agriculture. The division's early focus was the study of the role of birds in controlling insect pests and the geographical distribution of animal and plant species throughout the nation. Before the end of the nineteenth century, the expanding division was renamed the Bureau of Biological Survey.

Through merger and reorganization, the Commission on Fish and Fisheries and the Bureau of Biological Survey eventually became the Fish and Wildlife Service (FWS) of the Department of the Interior in 1940. During this evolution, a variety of federal laws were passed that strongly influenced the direction and authority of the FWS over environmental resources. The establishment of wildlife refuges was begun under President Benjamin Harrison. In 1900 the Lacey Act was passed, under which authority the federal government began to enforce laws regarding interstate and foreign commerce in wildlife. During Theodore Roosevelt's presidency, many more wildlife refuges were created from public domain lands. The Migratory Bird Treaty Act of 1918 significantly expanded the service's authority by involving two foreign governments, Canada and Mexico, in wildlife protection. The Migratory Bird Conservation Act of 1929, which established refuges for migratory waterfowl management, provided authority from which the National Wildlife Refuge System developed over the following decades.

The 1930's saw the passage of two important acts that ultimately would fall under the oversight of the FWS. The Migratory Bird Hunting and Conservation Stamp Act of 1934, often called the Duck Stamp Act, requires waterfowl hunters to purchase stamps, revenues from which are used to acquire and protect important wetlands. Over its history, the duck stamp program has generated enough revenue to protect approximately 1.8 million hectares (4.5 million acres) of waterfowl habitat. The Federal Aid in Wildlife Restoration Act of 1937 (commonly known as the Pittman-Robertson Act), which has been identified by some writers as the most significant wildlife act in U.S. history, established an excise tax on firearms, archery equipment, and ammunition used in hunting, as well as a manufacturers' tax on handguns. These tax dollars, which are specified for wildlife and fisheries, are apportioned to state wildlife agencies through the FWS.

One of the FWS's best-known responsibilities is coadministration of the Endangered Species Act (ESA), a duty it shares with the National Oceanic and Atmospheric Administration. Congress passed the ESA in 1973 to protect endangered plants and animals domestically and internationally. The FWS determines whether populations of freshwater and land species are threatened or endangered and designates critical habitats.

Other FWS responsibilities include operating the National Wildlife Refuge System (which encompasses some 38 million hectares, or 93 million acres), cooperating with state fish and game agencies and Native American tribal officials, assisting foreign governments with international conservation efforts, establishing university extension programs regarding fish and wildlife, and collecting, maintaining, and distributing statistical information on fisheries, wetlands, and estuaries. To carry out all of these duties, the FWS has field and regional offices across the country. The scope of the agency's activities reflects the wide range of environmental issues that affect fish and wildlife resources.

*Jerry E. Green*
*Updated by Karen N. Kähler*

## Further Reading

Clarke, Jeanne Nienaber, and Daniel McCool. *Staking Out the Terrain: Power and Performance Among Natural Resource Agencies.* 2d ed. Albany: State University of New York Press, 1996.

Hays, Samuel P. *A History of Environmental Politics Since 1945.* Pittsburgh: University of Pittsburgh Press, 2000.

U.S. Fish and Wildlife Service. *Shared Commitments to Conservation: 2007 Annual Financial Report of the U.S. Fish and Wildlife Service.* Washington, D.C.: U.S. Department of the Interior, 2007.

## Gila Wilderness Area

CATEGORY: Preservation and wilderness issues
DATE: Established in June, 1924
DEFINITION: Area of protected wilderness within the Gila National Forest
SIGNIFICANCE: Establishment of the Gila Wilderness Area, the first designated wilderness area in the world, was a harbinger of the national wilderness system later established under the Wilderness Act of 1964.

The Gila Wilderness Area, located in southwestern New Mexico, comprises more than 202,000 hectares (500,000 acres) of rugged country containing highly diverse landforms, plants, and animals. It includes the termini of both the Rocky Mountains to the north and the Sierra Madre range to the south. The Chihuahuan and Sonoran deserts also reach into the Gila Wilderness Area, contributing to its high biodiversity.

The designation of this region as the world's first official wilderness area grew out of the nineteenth century conservation movement in the United States. As the United States became increasingly settled, some people grew concerned with the need to preserve the land's natural beauty. This led to the establishment of the first national park, Yellowstone, in 1872. At the same time, however, plans were being formulated to make the park and similar areas readily accessible to visitors. About thirty years later, a number of observers began to realize that some wildlands needed to be preserved in a manner devoid of roads and other amenities for visitors.

In 1899 President William McKinley set aside the Gila River Forest Reserve. Seven years later, the Gila National Forest was established. In the early 1920's threats to the area by developers led Aldo Leopold, then a young forester in the Southwest District of the U.S. Forest Service, to awaken the American public to the need to save wildlands. An avid hunter, Leopold observed that wildlands and wildlife habitats were shrinking, and he blamed these losses on road building. Upon his arrival in the Southwest in 1909, he had identified six roadless backcountry areas—each more than 202,000 hectares in size—in the region's national forests. By 1921, only one, at the headwaters of the Gila River, remained without roads.

Passage of the Federal Highway Act in 1922 threatened to bring more road building and tourism, leading Leopold to fear the loss of this last roadless area. As assistant district forester, he led a movement to establish the Gila headwaters area as the first officially designated wilderness, to keep it roadless and without buildings or artificial trails. Persuaded by his arguments, the district forester designated some 305,500 hectares (755,000 acres) as the Gila Wilderness Area in June, 1924—a harbinger of the national wilderness system later established under the Wilderness Act of 1964.

Under the 1924 U.S. Forest Service wilderness designation, hunting was encouraged in the Gila Wilderness Area. Predator control then led to deer overpopulation. In 1933 the Forest Service responded by rebuilding an abandoned wagon road through the wilderness to provide easier access for hunters. This resulted in division of the wilderness. After some modifications in the laws and name changes in the intervening years, the New Mexico Wilderness Act of 1980 designated approximately 226,000 hectares (558,000 acres) west of the road as the Gila National Wilderness and about 81,000 hectares (200,000 acres) east of the road as the Aldo Leopold National Wilderness. These areas together exceed the size of the original Gila Wilderness Area.

*Jane F. Hill*

## Further Reading

Huggard, Christopher J. "America's First Wilderness Area: Aldo Leopold, the Forest Service, and the Gila of New Mexico, 1924-1980." In *Forests Under Fire: A Century of Ecosystem Mismanagement in the Southwest*, edited by Christopher J. Huggard and Arthur R. Gómez. Tucson: University of Arizona Press, 2001.

Lewis, Michael, ed. *American Wilderness: A New History*. New York: Oxford University Press, 2007.

Lorbiecki, Marybeth. *Aldo Leopold: A Fierce Green Fire*. 1996. Reprint. Guilford, Conn.: Globe Pequot Press, 2005.

## Glen Canyon Dam

CATEGORY: Preservation and wilderness issues
IDENTIFICATION: Large hydroelectric dam built on the Colorado River in northeastern Arizona
DATES: Construction begun in 1956; reservoir filled in 1983
SIGNIFICANCE: Construction of the Glen Canyon Dam flooded the pristine, wild Glen Canyon and created Lake Powell. Downstream, the dam's effects on water temperature and sediment load affected the ecology of the Grand Canyon over time. Flow regulation experiments begun during the late 1990's have sought to reverse some of the dam's adverse effects.

The U.S. Congress authorized the building of the Glen Canyon Dam in 1956 to provide water storage by creating the second-largest human-made reservoir in the United States, Lake Powell. Construction of the dam began in 1956, and the lake reached its full mark in 1983. The dam and reservoir generate hydroelectric power, provide a recreation area, and decrease siltation in the downstream Lake Mead reservoir, which is formed by Hoover (Boulder) Dam. The combined storage behind the Glen Canyon and Hoover dams manages the flow of water in the Colorado River to California, Arizona, Nevada, and Mexico. Agriculture consumes approximately 85 percent of the Colorado River water, with most of the remainder going to urban areas in Southern California; Phoenix, Arizona; and Las Vegas, Nevada. Glen Canyon Dam, with power plants, cost $272 million to build. Although the plants were designed to generate up to 1,300 megawatts of electric power annually, environmental damage to the riparian (shoreline) ecosystem downstream in the Grand Canyon proved too great at this rate of electricity production, and the government limited production to fewer than 800 megawatts annually.

When construction on the dam began, Glen Canyon was considered extremely remote and was rarely visited. The Sierra Club initially fought the project but agreed to end its resistance in a compromise agreement with the U.S. Bureau of Reclamation that stopped construction of two other dams in Utah. The environmental organization's leaders soon regretted their decision when they learned of the spectacular beauty of Glen Canyon. Sierra Club executive director David Brower teamed with photographer Eliot Porter to create a beautifully illustrated book titled *The Place No One Knew* (1963), which featured photographs of Glen Canyon before the dam. This book launched a Sierra Club policy of publishing illustrated books on outstanding natural areas to try to ensure that Americans would sacrifice no other spectacular areas because they did not know of their beauty. In 1996 the Sierra Club called on the federal government to drain Lake Powell and abandon the dam.

Many people view the Glen Canyon Dam as an icon of environmental destruction. The central theme of Edward Abbey's landmark novel on ecotage, *The Monkey Wrench Gang* (1975), involves plotting to destroy the dam. The radical preservationist group Earth First! leaped into national prominence during the early 1980's when members unfurled a 300-foot length of black plastic from the top of the dam that gave the appearance of a giant crack descending down the front of the structure. Brower suggested that the lake be drained but the dam left "as a tourist attraction, like the Pyramids, with passers-by wondering how humanity ever built it, and why."

### IMPACTS

The most obvious postdam change was the flooding of a spectacular canyon and its tributaries, along with the accompanying loss of hiking and river-running experiences. The entire ecosystem of the inundated area, 653 square kilometers (252 square miles), was destroyed and replaced with a lake community. The reservoir water altered the color of the adjacent red sandstone rocks, resulting in a prominent white "bathtub ring" along the shore most of the year.

The reservoir behind the Glen Canyon Dam caused several other changes. Because the reservoir is located in the desert, an estimated 74,000 hectare-meters (1 hectare of water, 1 meter deep) of water (600,000 acre-feet) are lost to evaporation each year. Annually, more than 43,000 hectare-meters (350,000 acre-feet) of water seep into the surrounding rock strata. The river loses approximately 8 percent of its volume in Lake Powell. An additional concern is the

*Glen Canyon Dam.* (©Dreamstime.com)

rate of siltation. Silt and sand carried by the Colorado River settle onto the reservoir floor, and more than 123,000 hectare-meters (1 million acre-feet) of water storage have been loss to siltation.

Dramatic changes to the downstream riparian environment in the Grand Canyon are less apparent. Prior to dam construction, muddy river temperatures seasonally ranged from 26 degrees Celsius (80 degrees Fahrenheit) to nearly freezing. The dam releases water from the depths of the reservoir, and the river below the dam is approximately 9 degrees Celsius (48 degrees Fahrenheit). This water is also more saline than it was before the dam was erected, as evaporation allows the natural salts in the water to become more concentrated. In addition, the river below the dam bears only 15 percent of the sediment and nutrient load that it used to carry when it ran free. These changes have affected the aquatic fauna and flora of the Grand Canyon's Colorado River. Several species have become extinct, and many nonnative or previously sparse species have flourished.

Releasing water from the dam to meet hydroelectric needs eliminated large floods but caused a daily tide as more water was released during afternoon, high-power-demand times and less flowed through the dam during the low-demand nights. The resulting tide stressed aquatic organisms downstream and caused rapid beach and sandbar erosion. Lack of natural flooding prevented rebuilding of the beaches and sandbars. The rapid loss of sandy deposits along the river in the Grand Canyon brought about ecosystem changes and had negative impacts on the recreational experiences of customers of the multimillion-dollar rafting industry.

### New Approaches

During the 1980's and 1990's, the U.S. Department of the Interior undertook an extensive review of the impacts of the Glen Canyon Dam that resulted in an environmental impact study. The major outcome of the $70 million study was a revised approach to the way water was released from the dam. Extreme and rapid daily fluctuations in river level to meet power needs were eliminated. Also, in the spring of 1996 the dam released an experimental flood with a slow rise to modest flood levels, followed by a gradual lowering of the river. The release was designed to scour the river channels and rebuild the beaches, as happens in a natural flood. This flow management measure succeeded in redistributing existing downstream sediments without contributing significant quantities of new sediment, but its benefits proved to be short-term. Other controlled flood experiments conducted in 2004 and 2008 also yielded transient benefits.

During the late 1990's, serious discussion began regarding the Sierra Club's call to decommission the dam, drain Lake Powell, and restore Glen Canyon. If such a plan were to be carried out, it would take the reservoir fifteen to twenty years to drain. According to an assessment issued in 2000 by the Glen Canyon Institute, the dam could continue to generate hydroelectric power for ten to fifteen years while the lake drained. Opponents of canyon restoration assert that decommissioning the dam would be too detrimental to water supplies, power generation, recreation, local economies dependent on tourism and the hydroelectric industry, and the ecosystems that have developed in the decades since the dam was built. Whether Lake Powell is maintained or drained, either choice would have a complex array of beneficial and adverse environmental effects on an extensive area upstream and downstream from Glen Canyon Dam.

*Louise D. Hose*
*Updated by Karen N. Kähler*

FURTHER READING

Carothers, Steven W., and Bryan T. Brown. *The Colorado River Through Grand Canyon: Natural History and Human Change.* Tucson: University of Arizona Press, 1991.

Farmer, Jared. *Glen Canyon Dammed: Inventing Lake Powell and the Canyon Country.* Tucson: University of Arizona Press, 1999.

Gloss, Steven P., Jeffrey E. Lovich, and Theodore S. Melis, eds. *The State of the Colorado River Ecosystem in Grand Canyon: A Report of the Grand Canyon Monitoring and Research Center, 1991-2004.* Reston, Va.: U.S. Geological Survey, 2005.

Lowry, William R. *Dam Politics: Restoring America's Rivers.* Washington, D.C.: Georgetown University Press, 2003.

McPhee, John. *Encounters with the Archdruid.* 1971. Reprint. New York: Farrar, Straus and Giroux, 2000.

Martin, Russell. *A Story That Stands Like a Dam: Glen Canyon and the Struggle for the Soul of the West.* Salt Lake City: University of Utah Press, 1999.

Parks, Timothy L. *Glen Canyon Dam.* Charleston, S.C.: Arcadia, 2004.

Porter, Eliot. *The Place No One Knew: Glen Canyon on the Colorado.* Edited by David Brower. Commemorative ed. Layton, Utah: Gibbs Smith, 2000.

Powell, James Lawrence. *Dead Pool: Lake Powell, Global Warming, and the Future of Water in the West.* Berkeley: University of California Press, 2008.

## Global ReLeaf

CATEGORIES: Organizations and agencies; activism and advocacy; forests and plants; preservation and wilderness issues
IDENTIFICATION: Conservation program that focuses on the planting of trees and the protection of forestlands from overdevelopment and pollution
DATE: Initiated on October 12, 1988
SIGNIFICANCE: Global ReLeaf has successfully planted millions of trees while also working to educate the public about the importance of the world's forests for the planet's environmental health.

In the United States, Global ReLeaf is the education and action program of the organization American Forests (formerly known as the American Forestry Association), which was founded in 1875 and is the nation's oldest national nonprofit conservation organization. Global ReLeaf sponsors educational programs to show the benefits of trees and forests for the environment and for the enhancement of people's lives. These programs highlight the value of trees for filtering air and water, sheltering and feeding wildlife, absorbing greenhouse gases, and reducing the runoff of polluted soil into rivers and streams. The program also provides funding, from private and corporate donations, for tree-planting projects across the United States.

Global ReLeaf's activities include ecosystem restoration projects called Global ReLeaf Forests, which involve the planting of trees, typically on public or private land that was once forested but has been cleared by wildfires, hurricanes, tornadoes, insects, or other natural occurrences; by developers; or by unintentional human interference, such as the spread of accidentally introduced exotic species. Program personnel work with local groups to ensure that the new trees are native to the area and that they are properly planted and maintained.

From the time of its initiation in 1988 through 2010, Global ReLeaf planted more than thirty million trees during more than six hundred projects, aiming toward a goal of one hundred million trees planted by 2020. In 2010 alone, the organization planted more than four million trees in fourteen U.S. states and ten countries around the world.

Global ReLeaf has been successful in part because it works with governmental agencies, local organizations, and large corporations to make it easy for individual citizens to participate. Through extensive advertising and publicity, and through a colorful presence on the Internet, American Forests has encouraged donations of as little as ten dollars, with one tree being planted for each dollar received. In partnership with a major breakfast cereal, Global ReLeaf Kids supported a project to plant trees in rain forests in the Philippines and Hawaii.

Other organizations throughout the world have also used the name Global ReLeaf. One prominent group based in Slovakia in Eastern Europe was the former Slovak Union of Nature and Landscape Conservation, now called the Global ReLeaf Foundation. Like American Forests, this organization sponsors educational programs and prepares school curriculum materials, especially about the dangers of pollution and overdevelopment.

*Cynthia A. Bily*

FURTHER READING

Cohen, Shaul E. "American Forests: Planting the Future." In *Planting Nature: Trees and the Manipulation of Environmental Stewardship.* Berkeley: University of California Press, 2004.

Gray, Gerald J., Maia J. Enzer, and Jonathan Kusel, eds. *Understanding Community-Based Forest Ecosystem Management.* Binghamton, N.Y.: Haworth Press, 2001.

## Grand Canyon

CATEGORIES: Places; preservation and wilderness issues
IDENTIFICATION: Deep gorge created by the Colorado River in northern Arizona
SIGNIFICANCE: The popularity of the Grand Canyon as a tourist destination has contributed to a number of environmental problems in and around the canyon itself. These problems range from air and noise pollution to issues related to the scarcity of water resources.

The Grand Canyon is a deep, 450-kilometer-long (280-mile) segment of the Colorado River and its tributary canyons in northern Arizona. Grand Canyon National Park, one of the most heavily visited national parks, was established by President Theodore Roosevelt in 1919 and has been designated a World Heritage Site. The parts of the Grand Canyon that lie outside the park's boundaries are managed by the Hualapai and Navajo tribal councils.

The arid climate of the Grand Canyon influences every resource in the region. Well-exposed rock layers reveal more than 1.8 billion years of the earth's history. The arid climate preserves ancient human and animal remains, including those of many extinct animals that lived more than ten thousand years ago. Cliff dwellings, human artifacts, and old adobe buildings represent habitation dating back more than four thousand years. Grand Canyon tourists, however, often focus

---

### The Goals of American Forests

*American Forests, the organization that conducts the Global ReLeaf program, describes the organization's mission, vision, and strategy as follows:*

**Mission:** Our mission is to grow a healthier world.
**Vision:** Our vision is to have healthy forest ecosystems for every community.
**Strategy:** Our strategy for achieving the mission is to provide action opportunities to targeted audiences to enable them to improve their environment with trees. We do this by using the best science to identify conservation issues, then develop and market practical solutions that individuals and groups can apply. American Forests' targeted audiences are individuals, community groups, government at all levels, educators, and businesses.

most on the canyon's scenic grandeur and beautiful views. People who hike into the canyon commonly feel overwhelmed by the immensity of the chasm.

### Ecosystems

Grand Canyon National Park, with more than 2,000 meters (6,500 feet) of relief, contains many ecosystems. The South Rim, a flat plateau, has an elevation of 2,100 meters (7,000 feet). The North Rim is an even higher plateau, 2,400 meters (8,000 feet) above sea level. Coniferous forests cover both rims and provide homes to deer, squirrels, and mountain lions. No streams cross the plateaus, as water from rain and snowmelt immediately flows underground in the karst terrains.

The coniferous forests extend down into the canyon, transitioning into an arid environment at lower elevations. Desert plants—such as cacti, acacia, mesquite, brittle bush, ocotillo, rabbitbrush, and agave—grow on the walls of the Grand Canyon. Desert bighorn sheep, lizards, snakes, skunks, and mice populate the slopes and side canyons. Water is scarce, particularly on the south side of the Colorado River. Cottonwood, ash, and redbud trees—as well as ferns, columbine, and other water-loving plants and animals—cluster around small seeps throughout the canyon and larger karst springs in some tributary canyons on the north side.

Another distinct ecological zone in the Grand Canyon is the riparian (riverside) habitat along the Colorado River at the bottom. Willow, arrowweed, and exotic tamarisk line the riverbanks. Otters, beavers, muskrats, fish, and other aquatic organisms call the Colorado River home.

*Visitors view the Grand Canyon at sunset from Mather Point on the South Rim after a dusting of snow.* (NPS)

## Environmental Problems

After visiting part of the Grand Canyon in 1858, Lieutenant Joseph C. Ives wrote, "The region is, of course, altogether valueless. . . . It seems intended by nature that the Colorado River . . . shall be forever unvisited and undisturbed." If his prediction had come true, the Grand Canyon would not be facing the many environmental problems that now threaten it.

More than five million people visit Grand Canyon National Park each year. Support facilities for these visitors (campgrounds, hotels, shops, toilets) require water and sewage treatment, yet no water is available on either rim. Drilling to the closest aquifer, several thousands of feet underground, would be extremely costly, and so all water used in the park comes from a cave about halfway up the north side of the canyon. A transcanyon pipeline and associated pumphouses lift the water to two places on the South Rim and one location on the North Rim. The purity and quantity of this modest stream are critical to keeping the Grand Canyon open to visitors. Occasionally, the water's impurities exceed state-set limits, and the park is forced to truck in water at tremendous expense. The numbers of visitors to the park and further development there will remain strictly limited unless other sources of water are developed.

The popularity of the Grand Canyon brings other problems as well. In an attempt to provide each visitor with a high-quality wilderness experience, the park requires overnight campers to register for permits and limits the number of visitors. In popular backcountry areas, campers are restricted to designated campgrounds that have solar-powered compost toilets. Because the desert recovers slowly from erosion, the park rangers enforce strict rules concerning vandalism and the cutting of switchbacks. They warn hikers to treat archaeological sites with respect. As in all national parks, visitors are forbidden to take any archaeological or historical materials, rocks, animals, or plants.

Because many visitors choose to see the park from the air, the presence of helicopters and low-flying airplanes has created a volatile issue. Many hikers have expressed dismay, and even anger, at the noise that accompanies the flights overhead, noting that it hinders their ability to enjoy the canyon's beauty and grandeur. The government struggles to balance the demands of these two different user groups by strictly controlling the routes and heights of overflights.

> ### Powell on the Great and Unknown Grand Canyon
>
> *John Wesley Powell combined into one narrative the story of his two trips into the Grand Canyon and down the Colorado River with his survey crew. In this excerpt from his 1875 book,* Exploration of the Colorado River of the West and Its Tributaries, *Powell introduces the reader to the start of the journey "down the Great Unknown."*
>
> We are now ready to start on our way down the Great Unknown. Our boats, tied to a common stake, chafe each other as they are tossed by the fretful river. They ride high and buoyant, for their loads are lighter than we could desire. We have but a month's rations remaining. The flour has been resifted through the mosquito-net sieve; the spoiled bacon has been dried and the worst of it boiled; the few pounds of dried apples have been spread in the sun and reshrunken to their normal bulk. The sugar has all melted and gone its way down the river. But we have a large sack of coffee. The lightening of the boats has this advantage: they will ride the waves better and we shall have but little to carry when we make a portage.
>
> We are three quarters of a mile in the depths of the earth, and the great river shrinks into insignificance as it dashes its angry waves against the walls and cliffs that rise to the world above; the waves are but puny ripples, and we but pigmies, running up and down the sands or lost among the boulders.
>
> We have an unknown distance yet to run, and unknown river to explore. . . .
>
> *Source:* Excerpted in *The Wilderness Reader,* edited by Frank Bergon (Reno: University of Nevada Press, 1980), p. 152.

Once known for being spectacularly clear, the air of the Grand Canyon is now occasionally marred by pollution from nearby coal-burning power plants. Since the 1970's, battles have raged between environmentalists concerned with preserving air quality in the area and those interested in developing the region's abundant coal supplies.

The most dramatic and contentious environmental issues in the Grand Canyon involve the Colorado River. The greatest change to the riparian zone resulted from construction of the Glen Canyon Dam, which has controlled the Colorado River flow through the Grand Canyon since 1966. The dam eliminated large floods, but the maintenance of more consistent flows throughout the year has had a severe impact on the riparian ecosystem, including elimination of many beaches and a general increase in vegetation (including nonnative species of plants) and wildlife. The dam has also altered water temperature and clarity. Predam water temperatures in the river fluctuated from 26 degrees Celsius (80 degrees Fahrenheit) in summer to nearly freezing in winter. Today, the river temperature remains constant at 7 degrees Celsius (46 degrees Fahrenheit). Once noted for its load of sediment, the river below the dam is now clear. The combined changes in water temperature and clarity have had dramatic effects on the canyon's aquatic ecology, resulting in the loss of many plant and fish species and the proliferation of nonnative carp and trout.

*Louise D. Hose*

FURTHER READING

Finkbine, Bob. *Smoke That Roars: The Grand Canyon and the Creation of a New Century*. Phoenix: Atwell, 2002.

Houk, Rose. *An Introduction to Grand Canyon Ecology*. Grand Canyon, Ariz.: Grand Canyon Association, 1996.

McPhee, John. *Encounters with the Archdruid*. 1971. Reprint. New York: Farrar, Straus and Giroux, 2000.

Webb, Robert. *A Century of Change: Rephotography of the 1889-90 Stanton Expedition*. Tucson: University of Arizona Press, 1996.

## Grand Coulee Dam

CATEGORIES: Preservation and wilderness issues; animals and endangered species
IDENTIFICATION: Dam located on the Columbia River west of Spokane, Washington
DATE: Completed in 1941

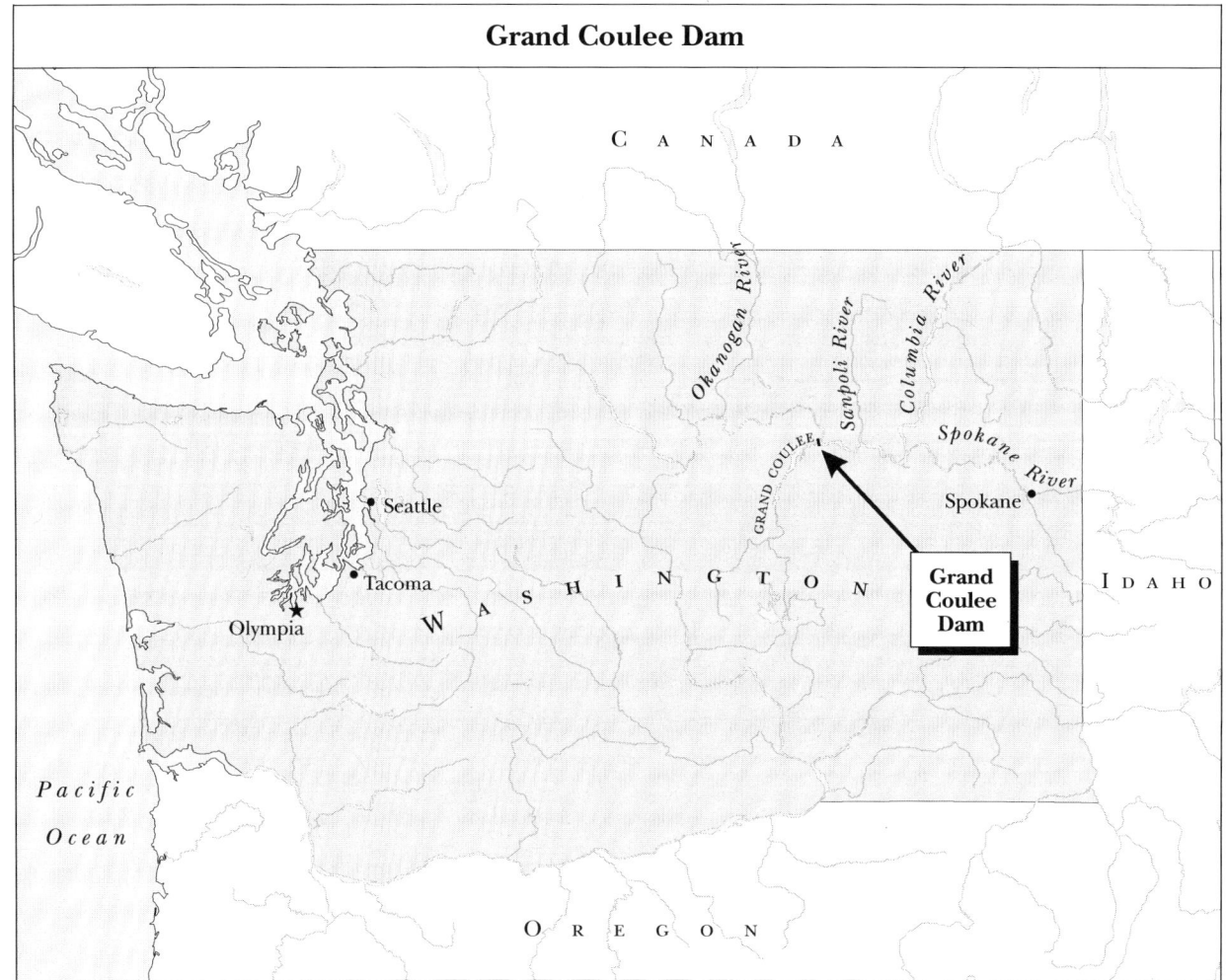

SIGNIFICANCE: The environmental impacts of the Grand Coulee Dam include a dramatic decrease in the salmon population in the Columbia River, once the largest natural salmon hatchery in the world.

The multipurpose, 168-meter-high (550-foot-high) Grand Coulee Dam, one of the largest concrete structures in the world, provides downstream flood control, irrigation, and hydroelectricity. The facility delivers irrigation water to more than 223,000 hectares (550,000 acres) of agricultural lands. Its associated electrical power production facilities are the largest in North America. President Franklin D. Roosevelt initiated construction of the dam in 1933 under the Public Works Administration of the National Industrial Recovery Act, which Congress authorized for emergency projects to relieve unemployment during the Great Depression. Electricity was first generated at the dam in 1941. Placing a dam across the fourth-largest river in North America created a 243-kilometer-long (151-mile-long) reservoir, Franklin D. Roosevelt Lake, which affords fishing and water sports to visitors.

Salmon ladders were built adjacent to downstream dams to allow the annual Columbia River salmon spawning migration to continue, but the Grand Coulee Dam is too high to accommodate this dam-passing technique. At the time of construction, an elevator to lift the fish up and into the reservoir was considered but rejected because too many fish would die in the process. Instead, more than 1,600 kilometers (1,000 miles) of the world's most prolific salmon breeding river was eliminated upstream of the dam. Many miles of the affected river flow through Canada, but the Canadian government expressed a lack of concern about the potential loss, stating that no commercial salmon fisheries existed along the river in Canada.

In the spring of 1939 the U.S. Bureau of Reclamation captured chinook and blueback salmon downstream from the Grand Coulee Dam site and transported them to spawning grounds in both tributaries and the main river upstream from the dam. The fish that survived the truck ride and transplant spawned. They and their offspring returned over the dam spillways and through the turbines, causing high mortality rates. The following year, captured fish were transferred to a hatchery for spawning. The young fish were released in downstream tributaries with the hope that they would return to the new spawning sites the following year. In 1943 the Bureau of Reclamation declared the salmon relocation program successful and turned the efforts over to the U.S. Fish and Wildlife Service.

The salmon population in the Columbia River, once the largest natural salmon hatchery in the world, dropped dramatically with construction of the Grand Coulee and other dams. At the downstream Rock Island Dam, more than fifty thousand salmon passed through ladders to bypass the dam in 1933. A 1942 census at the same site counted only a little more than seven thousand salmon. Improved understanding and accommodation of salmon diet and environmental needs has helped the Columbia River populace grow, but it remains far smaller than the predam population.

*Louise D. Hose*

FURTHER READING

McCully, Patrick. *Silenced Rivers: The Ecology and Politics of Large Dams.* Enlarged ed. London: Zed Books, 2001.

Moss, Brian. *Ecology of Fresh Waters: A View for the Twenty-first Century.* 4th ed. Hoboken, N.J.: Wiley-Blackwell, 2010.

Palmer, Tim. *Lifelines: The Case for River Conservation.* 2d ed. Lanham, Md.: Rowman & Littlefield, 2004.

# Great Swamp National Wildlife Refuge

CATEGORIES: Places; animals and endangered species; preservation and wilderness issues
IDENTIFICATION: Federally protected wildlife habitat in Morris County, New Jersey
DATE: Established on November 3, 1960
SIGNIFICANCE: The Great Swamp National Wildlife Refuge, which was established largely as the result of a grassroots effort, provides an important nesting and feeding habitat for migratory birds.

The Great Swamp National Wildlife Refuge in New Jersey occupies a 3,116-hectare (7,700-acre) region of bottomland hardwood swamps and mixed hardwood forests containing cattail marshes, grasslands, ponds, and streams. Its beginnings were established in 1960, and it was designated a National Natural Landmark in 1966. An estimated 300,000 people visit the refuge each year. The refuge supports approximately 240 species of birds, 39 species of reptiles and amphibians, 29 species of fish, 33 species of mammals, and 600 species of plants, 215 of which are wildflowers. Of those species, more than two dozen have been designated as threatened or endangered by the state of New Jersey, including the bog turtle, the wood turtle, and the blue-spotted salamander.

In 1959 the New York-New Jersey Port Authority identified a 4,047-hectare (10,000-acre) area in rural New Jersey as the site of a new airport to serve the New York City metropolitan area. The proposed site would cover twice the area of what was then known as Idlewild Airport (later renamed John F. Kennedy International). When the *Newark Evening News* broke the story about the proposed airport, citizens, politicians, and conservationists banded together to fight the project. Citizens objected to the Port Authority's expansion plans for a multitude of reasons: destruction of homes and businesses, unacceptable noise levels from the new airport, traffic, and contamination of underground water supplies.

Fourteen volunteer groups joined forces as the Jersey Jetport Site Association (JJSA). The JJSA fought the airport expansion on a variety of fronts: political, legal, and economic. Most important was the group's ability to influence public opinion. Some of the same people also became involved with the North American Wildlife Foundation (NAWF), which purchases threatened lands and holds them for future government purchase or donates the property outright. Through the efforts of the NAWF and the Bureau of Sport Fisheries and Wildlife, the U.S. Department of the Interior agreed to grant the area wildlife refuge status if the NAWF could raise the funds to purchase 1,214 hectares (3,000 acres). The NAWF acquired the first 405 hectares (1,000 acres) in 1960 and turned them over to the U.S. Fish and Wildlife Service later that same year, and on an November 3, 1960, an act of Congress established as a park the first part of what would eventually become the Great Swamp National Wildlife Refuge.

In 1962 public hearings began in New Jersey on a bill that would prohibit airport construction in seven northern counties—including the Port Authority site in Morris County. The bill passed by a wide margin but was vetoed by New Jersey governor Robert Meyner, who declared it unconstitutional. Governor Meyner lost his bid for reelection and was replaced by Richard Hughes, who supported the bill and the refuge. By 1963 the state had even provided $25,000 to purchase additional land, which was added in 1964. Additional parcels of land were appended to the existing refuge throughout the following years, until by 1990 the Great Swamp National Wildlife Refuge consisted of more than 2,800 hectares (7,000 acres) of land.

*P. S. Ramsey*

FURTHER READING

Dawson, Chad P., and John C. Hendee. *Wilderness Management: Stewardship and Protection of Resources and Values.* 4th ed. Boulder, Colo.: WILD Foundation, 2009.

Richman, Steven M. *The Great Swamp: New Jersey's Natural Treasure.* Atglen, Pa.: Schiffer, 2008.

## Green Plan

CATEGORY: Preservation and wilderness issues
IDENTIFICATION: Canadian national strategy to create a cleaner, safer, and healthier environment along with a sound economy
DATE: Initiated on December 11, 1990
SIGNIFICANCE: By establishing a series of sustainable development goals and then measuring progress toward those goals and mobilizing collective, nationwide efforts, Canada's Green Plan has provided a model for a national approach to environmental management.

After many years of extensive consultations with Canadians representing government, business, interest groups, and the public, Canada's Department of the Environment launched its internationally acclaimed Green Plan in December, 1990. The overall goal of the plan was to ensure that current and future generations would enjoy a safe, healthy environment and a sound economy. Although the Green Plan focused on a wide range of environmental issues, it also incorporated the fundamentals of sustainable development into all aspects of decision making at all levels of society.

Since the Green Plan was an umbrella document, many of the details were left to work out during implementation. Many programs were initiated that affected various aspects of the lives of all Canadians, including the air they breathed, the water they drank, and the food they ate. For example, numerous full assessments of priority toxic substances were performed, and in 1992 the number of full-time Canadian environmental inspectors and investigators was increased from forty-nine to seventy.

The Green Plan established a series of sustainable development goals for Canadians that now serve as benchmarks for measuring progress and mobilizing collective, nationwide efforts. The first goal is to ensure that current citizens and future generations have clean air, water, and land, all of which are essential to sustaining human and environmental health. Steps toward achieving this goal include the reduction of ground-level ozone (smog) to below the threshold of adverse health effects, and the reduction of Canada's generation of waste by 50 percent.

The second goal is the sustainable use of renewable resources, which involves shifting forest management from sustained yield to sustainable development. Some of the key areas for decision making are those of harvesting practices (particularly in old-growth forests), reforestation, and the use of forest pesticides. Answers are being provided through the creation of a network of model forests and the creation of Tree Plan Canada.

The third goal is the protection of special spaces and species. The Canadian government has set aside 12 percent of the country as protected space for parks, wildlife areas, and ecological reserves. Similarly, the fourth goal focuses on preserving and enhancing the integrity, health, biodiversity, and productivity of Canada's Arctic ecosystems. Waste cleanups and assessments have been carried out at numerous sites in the Yukon and the Northwest Territories.

The fifth goal of the Green Plan is to enhance Canada's commitment to global environmental security. For example, plans were implemented to phase out the use of human-made chlorofluorocarbons (CFCs), methyl chloroform, and other major ozone-depleting substances by the year 2000. In addition, national emissions of carbon dioxide and other greenhouse gases were to be stabilized at 1990 levels. The Green Plan also includes the goals of minimizing the impact of environmental emergencies and making environmentally responsible decisions. The Canadian government is accomplishing these goals by implementing plans for quick, effective responses to environmental emergencies and by providing accurate, accessible information about the environment to all Canadians.

*Alvin K. Benson*

FURTHER READING

Boyd, David R. *Unnatural Law: Rethinking Canadian Environmental Law and Policy.* Vancouver: University of British Columbia Press, 2003.

Dwivedi, O. P., et al. "The Canadian Political System and the Environment." In *Sustainable Development and Canada: National and International Perspectives.* Orchard Park, N.Y.: Broadview Press, 2001.

## Greenbelts

CATEGORIES: Land and land use; preservation and wilderness issues

DEFINITION: Tracts of open space preserved to control urban growth patterns

SIGNIFICANCE: Greenbelts provide numerous environmental, social, and economic benefits to the areas that surround them.

Throughout the twentieth century and into the twenty-first, developed nations have urbanized at ever-increasing rates. As a result of the automobile, the United States and Canada in particular have been subjected to uncontrolled growth, resulting in the phenomenon referred to as urban sprawl. As development moves outward from a central city, prime agricultural and forested lands are converted to more intensive uses, resulting in a significant loss of wildlife and plant habitats. This decrease in natural areas also leads to a subsequent degradation of air and water quality.

The concept of creating greenbelts, or greenways, developed as a grassroots response to address these problems. With limited public funds for open-space preservation, greenbelt proponents have focused attention on "leftover" or abandoned lands. These parcels are often found along ridgelines and streams, areas that are too steep or too wet for development. Abandoned railroad and utility rights-of-way have become important potential resources as well. All of these areas have common physical characteristics: They are long, thin tracts of land that relate to the topography, threading through land more suitable for development.

Greenbelts, as linear open-space corridors, provide several important benefits. First, they enable urban areas to retain their biodiversity. This is important for maintaining plant and animal habitats, as well as for establishing sources of protection for air and water quality. The natural corridors provide migration routes for species interchange. This movement of plant and wildlife along natural pathways is particularly significant, since it may determine the ability of some species to survive in these areas. Second, the retention of undeveloped, vegetated lands allows surface water to be returned naturally to the water table, minimizing surface runoff, erosion, and subsequent stream sedimentation.

Greenbelts offer many recreational opportunities as well. Most greenbelts include systems of trails that may link larger, more intensive recreational facilities or provide people with access to natural amenities from urban areas. By connecting different sorts of facilities, in essence creating a system or network of urban parks, greenbelts increase the aggregate benefit to the community. Because of the linear nature of greenbelts, they have more edge area than do other kinds of parks or open spaces. This characteristic maximizes the available open space and provides potential access to greater numbers of people.

The economic benefits of greenbelts are also significant. As leftover or derelict lands, suitable parcels may be purchased relatively inexpensively; thus minimum expenditure is often required for the development of a greenbelt system. In addition, the aesthetic improvement of the green edge provided by a greenbelt often enhances the value of adjacent properties.

*Steven B. McBride*

FURTHER READING

Amati, Marco, ed. *Urban Green Belts in the Twenty-first Century*. Burlington, Vt.: Ashgate, 2008.

Hough, Michael. *Cities and Natural Process: A Basis for Sustainability*. 2d ed. New York: Routledge, 2004.

# Habitat destruction

CATEGORIES: Preservation and wilderness issues; land and land use

DEFINITION: Degradation of a natural landscape so that it becomes functionally incapable of supporting its native species

SIGNIFICANCE: The destruction of habitat represents a pressing threat to global biodiversity, as habitat loss leads to species extinctions. In the twenty-first century most habitat destruction is occurring in developing nations, where overpopulation and poverty contribute to the need to convert forestland to agriculture.

Habitat destruction occurs when human beings remove or significantly alter the land or aquatic communities in which other animal species dwell. Certainly, habitat destruction is not always a negative practice; human civilization was built on the practice of altering land for human purposes, such as the burning of forest for pasture, logging for construction materials, the draining of marshland for development, and mining for resource extraction. In the twenty-first century, however, the density of human population on the earth, combined with modern industrial technology, means that humans' alterations of the natural landscape greatly exceed the ability to render land productive for human needs. Further, the most biodiverse ecoregions of the world, tropical rain forests and coral reefs, have seen rapid increases in habitat destruction. Tropical forests are down to about 1 billion hectares (2.5 billion acres) from the nearly 1.6 billion hectares (4 billion acres) they occupied approximately two hundred years ago. One-fifth of coral reefs have been destroyed, and another one-fifth have been severely degraded.

Habitat destruction is most often caused by expansion of agriculture. Planting crops or raising livestock requires wide expanses of bare soil or grassy plain. When suitable land is unavailable naturally, many land types can be converted to agricultural uses: Wetlands can be drained, forests can be logged, deserts can be irrigated. Although the productivity of such converted land may be relatively high, biodiversity and ecosystem functionality drop harshly. Modern agriculture is often monocultural (that is, devoted to a single type of crop), without topographical complexity, and devoid of any plants not being grown for either sale or consumption. Therefore, the available niches for native species are nearly all removed.

Other causes of habitat destruction include mining, ocean trawling, oil prospecting, urban sprawl, and infrastructure development. These practices directly destroy ecosystems, but other practices indirectly degrade surrounding ecosystems. Desertification is caused by overgrazing of livestock and excessive extraction of groundwater, which render the land unusable, usually affecting communities that already live in resource-impoverished landscapes. Deforestation on a small scale can cause ecosystem collapse by dividing a forest into fragments, rendering the land unfit to support animals with large ranges and plants with wide dispersal needs. Coral degradation is rarely caused by direct destruction; rather, coral is negatively affected by increases in water temperature and changes in water chemistry resulting from climate change and industrial pollution.

Depending on terrain characteristics and climate patterns, landscapes naturally acquire ecosystem types that are functional to their locations. When humans alter these land types for unnatural purposes, the consequences can be deleterious. Modern examples include the levy system in New Orleans, which replaced an extensive natural wetland buffer; when Hurricane Katrina hit the city in 2005, the levy system failed and the city was flooded. The devastation that resulted when Haiti was struck by a massive earthquake in 2010 was magnified by the high rate of deforestation, and therefore high rate of soil erosion, in that nation. Maintaining functional natural landscapes is increasingly seen as a priority, and restoration efforts are being developed to restore native habitat types even in urban areas.

### Developing Countries

Since the mid-twentieth century, most of the world's habitat destruction has taken place in developing countries, as developed countries have already exhausted their most accessible resources and altered much of their land for development. For example, nearly 50 percent of wetlands in the United States and 60-70 percent of European wetlands have been destroyed.

The main factors contributing to land alteration in the developing world are poverty, overpopulation, lack of sustainable technology, and adherence to cultural practices. For example, many communities in the developing world frequently cook with charcoal, as electricity and natural gas are in short supply. Charcoal is acquired through the burning of forestland, and the results are mass deforestation and air pollution. New technologies that do not require large monetary investments, such as solar ovens and permaculture techniques, are increasingly being offered as solutions to poor communities that rely on habitat destruction to survive.

Many farmers in the developing world are wary of foreign technology and are unwilling to risk the failure of a crop to adopt a new technology, even if it could result in increased production. Because of this wariness, education is needed to help farmers ease into the use of new technologies that can benefit them, and the environment, in the long term. For ecological conservation to be sustainable, it must be beneficial for both the farmers and the environment.

### Future Solutions

The ability of humans to alter the land for their own gain is one of the main reasons humans have been able to expand over the planet and live in nearly every climate. Since the dawn of large civilizations, humans have increasingly acquired the means to reap more and more from the environment, sometimes altering it so severely that they have eradicated whole communities of species. Mass extinctions occurred in prehistoric times during the early years of humanity, but the species that died off at that time were mostly those targeted for food. In contrast, in modern times entire ecosystems are destroyed for resources and agriculture, and this can only have increasingly negative effects on both the natural world and human habitations.

In order to measure the direct importance for humans of many landscapes, scientists have defined the ecosystem services provided by the landscapes. These services, such as erosion prevention, storm buffering, soil productivity, and wildfire prevention, are often given monetary equivalents to introduce them into a system of economics.

Other approaches to reducing habitat destruction include the use of modern techniques to increase agricultural productivity that can reduce the need to clear more land for farming. Planting native species, creating tree-shaded spaces, and increasing topographical heterogeneity on agricultural land can increase biodiversity without having a large impact on production. In addition, the incorporation of natural areas around cities to buffer weather, prevent erosion, and safeguard watersheds may save much money in repairs and provide protection against disasters.

*Jamie Michael Kass*

### Further Reading

Barbault, R., and S. D. Sastrapradja. "Generation, Maintenance, and Loss of Biodiversity." In *Global Biodiversity Assessment*, edited by V. H. Heywood. New York: Cambridge University Press, 1995.

Cincotta, Richard P., and Robert Engelman. *Nature's Place: Human Population Density and the Future of Bio-*

*logical Diversity*. Washington, D.C.: Population Action International, 2000.

Pullin, Andrew S. "Effects of Habitat Destruction." In *Conservation Biology*. New York: Cambridge University Press, 2002.

Tibbetts, John. "Louisiana's Wetlands: A Lesson in Nature Appreciation." *Environmental Health Perspectives* 114 (January, 2006): A40-A43.

## Hetch Hetchy Dam

CATEGORY: Preservation and wilderness issues
IDENTIFICATION: Part of a hydroelectric system built in Yosemite National Park to provide water and power to the city of San Francisco
DATE: Completed in 1923
SIGNIFICANCE: The controversy over O'Shaughnessy Dam's construction in Yosemite National Park during the early twentieth century was the first environmental issue argued on the national stage in the United States. Although environmentalists were unsuccessful in stopping the construction, they developed strategies and gathered support that became useful in later battles—including twenty-first century efforts to have the dam removed.

In 1890, eighteen years after the U.S. Congress named Yellowstone the first national park in the United States, Yosemite was named the second. Occupying some 3,100 square kilometers (1,200 square miles) in the Sierra Nevada in eastern California, the new park featured giant redwoods and sequoias and two great scenic mountain valleys less than 32 kilometers (20 miles) apart: Yosemite Valley (which technically remained under state control for several more years) and Hetch Hetchy Valley. Both valleys offered breathtaking wilderness: flowering meadows sur-

*View of the Hetch Hetchy Valley before it was dammed.* (Sierra Club Bulletin, 1908)

rounded by sheer cliffs of colorful granite punctuated by dramatic waterfalls. Writers such as John Muir, John Burroughs, and Mary Austin tramped through both valleys and the surrounding glacier-scoured mountains, bringing back descriptions of awe-inspiring beauty. Many thought that of the two, the oddly named Hetch Hetchy was the more beautiful. The 5.6 kilometers (3.5 miles) of the flat valley floor were traversed by a clear, clean river, and its granite walls were straight and steep. Because Yosemite's state control was less stringent than Hetch Hetchy's federal control, concessions and tourist businesses sprang up around Yosemite Valley, making it the more popular attraction.

San Francisco, 240 kilometers (150 miles) to the west, was one of the fastest-growing cities in the United States. As its population increased, so did its need for fresh water and electricity. At the time there was no public water supply, and San Franciscans were at the mercy of private companies that were not always responsive or responsible. By the beginning of the twentieth century, San Francisco's need for water was desperate. In 1901, under pressure from California legislators, the U.S. Congress passed the Right of Way Act, giving local governments the right to use national park lands for water projects such as dams and reservoirs if the projects were in the public interest. San Francisco officials wasted no time in declaring their intention to build a dam on the Tuolumne River at the narrow end of Hetch Hetchy Valley, flood the valley, and create a reservoir to supply the city with water. The very things that contributed to the valley's natural beauty—the flat valley floor, the steep cliffs, and the purity of the river—also made it an ideal spot for a reservoir.

When the city first applied for a right-of-way in 1903, the request was denied by U.S. secretary of the interior Ethan A. Hitchcock, who felt a dam was not in the public interest. After San Francisco's devastating earthquake and fire of 1906, the city's efforts intensified. Gifford Pinchot, chief of the U.S. Forest Service, was sympathetic to San Francisco's plan, and he urged the city to reapply for a right-of-way. It appeared that the new secretary of the interior, James Rudolph Garfield, was inclined to grant permission for the dam.

Immediately John Muir and the Sierra Club sprang into action to prevent the project. Muir wrote a personal letter to his old hiking companion, President Theodore Roosevelt, asking that other locations be

---

### Muir Opposes the Dam

*In his 1912 book* The Yosemite, *John Muir argued eloquently against the damming of the Tuolumne River in the Hetch Hetchy Valley:*

Hetch Hetchy Valley, far from being a plain, common, rock-bound meadow, as many who have not seen it seem to suppose, is a grand landscape garden, one of Nature's rarest and most precious mountain temples. As in Yosemite, the sublime rocks of its walls seem to glow with life, whether leaning back in repose or standing erect in thoughtful attitudes, giving welcome to storms and calms alike, their brows in the sky, their feet set in the groves and gay flowery meadows, while birds, bees, and butterflies help the river and waterfalls to stir all the air into music—things frail and fleeting and types of permanence meeting here and blending, just as they do in Yosemite, to draw her lovers into close and confiding communion with her.

Sad to say, this most precious and sublime feature of the Yosemite National Park, one of the greatest of all our natural resources for the uplifting joy and peace and health of the people, is in danger of being dammed and made into a reservoir to help supply San Francisco with water and light, thus flooding it from wall to wall and burying its gardens and groves one or two hundred feet deep. This grossly destructive commercial scheme has long been planned and urged (though water as pure and abundant can be got from sources outside of the peoples park, in a dozen different places), because of the comparative cheapness of the dam and of the territory which it is sought to divert from the great uses to which it was dedicated in the Act of 1890 establishing the Yosemite National Park.

The making of gardens and parks goes on with civilization all over the world, and they increase both in size and number as their value is recognized. Everybody needs beauty as well as bread, places to play in and pray in, where Nature may heal and cheer and give strength to body and soul alike. This natural beauty-hunger is made manifest in the little windowsill gardens of the poor, though perhaps only a geranium slip in a broken cup, as well as in the carefully tended rose and lily gardens of the rich, the thousands of spacious city parks and botanical gardens, and in our magnificent National parks—the Yellowstone, Yosemite, Sequoia, etc.—Nature's sublime wonderlands, the admiration and joy of the world.

*Source:* John Muir, *The Yosemite* (New York: Century, 1912).

developed instead. Roosevelt agreed that other rivers and dams should be exploited first but did not completely rule out the eventual flooding of Hetch Hetchy. Sierra Club members and others wrote letters to the editors of major newspapers on both coasts and garnered enough support for preserving Hetch Hetchy to stall the project in Congress.

### Passage of the Raker Act

For the next six years the issue was debated by the Congress and by the public in newspapers, newsletters, and public addresses. San Francisco's need for water was real, but many people argued that the need should be met in some way that would not destroy irreplaceable wilderness. Some argued that a reservoir in the valley would actually be more beautiful and attract more tourists to the area than the wild valley. For others, the issue was a matter of private versus public control of the city's water supply. Still others expressed the debate in terms of Pinchot's preservationism and the Sierra Club's conservationism.

Several congressional hearings on the right-of-way were held between 1908 and 1913. A brochure written by Muir in 1911 titled *Let Everyone Help to Save the Famous Hetch-Hetchy Valley and Stop the Commercial Destruction Which Threatens Our National Parks* included tips on lobbying Congress—a strategy that remains common for environmentalists today. Although conservationists attracted an impressive amount of support across the nation, they were ultimately defeated. In 1913 Congress passed the Raker Act, giving permission for the construction of a dam at Hetch Hetchy Valley.

O'Shaughnessy Dam—at the time the nation's largest concrete dam—was finished in 1923, and the associated water and power system was completed in 1934. Additional construction in 1938 raised the dam's height. The flooded valley never did attract many tourists, even during the late 1990's, when Yosemite Valley was so crowded that access by car was restricted. Water from the reservoir did help meet the needs of San Francisco, but only after control over its distribution was turned over to a private utility.

In 1987 Secretary of the Interior Donald P. Hodel proposed to study the removal of O'Shaughnessy Dam and the restoration of Hetch Hetchy Valley. State and federal studies concluded that the valley was more valuable as a water source than as a restored environment. During the early twenty-first century public interest in dam removal and valley restoration increased after the release of major reports by the Environmental Defense Fund (2004) and the organization Restore Hetch Hetchy (2005), along with two associated master's theses (2003 and 2004), all of which generally supported the concept of returning Hetch Hetchy to its original state. A 2006 California Resources Agency restoration study found the concept feasible, and the report on the research called for additional study. Although hundreds of small dams have been removed in the United States, no removal has ever been conducted for a dam as large as O'Shaughnessy. Restore Hetch Hetchy stated its hope of winning congressional approval for dam removal by December, 2014, the month marking the one hundredth anniversary of Muir's death.

*Cynthia A. Bily*
*Updated by Karen N. Kähler*

### Further Reading

Jones, Holway R. *John Muir and the Sierra Club: The Battle for Yosemite.* San Francisco: Sierra Club Books, 1965.

Righter, Robert W. *The Battle Over Hetch Hetchy: America's Most Controversial Dam and the Birth of Modern Environmentalism.* New York: Oxford University Press, 2005.

Simpson, John W. *Dam! Water, Power, Politics, and Preservation in Hetch Hetchy and Yosemite National Park.* New York: Pantheon Books, 2005.

# Indigenous peoples and nature preservation

CATEGORIES: Philosophy and ethics; preservation and wilderness issues

DEFINITION: Involvement of the native inhabitants of a region in the protection of the region's natural resources

SIGNIFICANCE: The role of indigenous peoples in nature preservation is a growing side of environmental awareness. While some people are practicing preservation by learning from indigenous peoples' traditional ecological knowledge, indigenous peoples elsewhere are fighting for recognition and political power to preserve their lands and cultures.

Nature preservation encompasses the protection of biodiversity, land, and culture. Nature reserves are areas set aside by governments to protect the species that inhabit those areas from harm or loss. Such reserves generally seek to protect natural areas from human disturbance to reduce biodiversity loss and prevent extinction of species and destruction of habitats. Protecting nature through preservation in this way is often not the best approach for managing natural resources, however; the reality is that ecosystems, and the biodiversity within them, move through natural changes, and trying to prevent these changes can have negative impacts on ecosystem health.

Many human activities threaten the biodiversity of the natural world, including logging, drilling, and industries that cause pollution. Another threat to the natural world is the fading of indigenous cultures that have long traditional relationships with the biodiversity of the areas in which they live. Indigenous peoples have evolved not with the practice of preservation but instead with the practices of adaptation and healthy use of natural resources. An important part of preserving biodiversity is to preserve the lands and cultures of the indigenous peoples who best recognize how to maintain healthy relationships with the environment.

## Traditional Ecological Knowledge

The accumulated knowledge of indigenous peoples regarding the environments they inhabit and their relationships with those environments is often referred to as traditional ecological knowledge, or TEK. Indigenous cultures around the world have a great accumulation of TEK through generations of familiarizing themselves with their home areas, and many scientists recognize the value of TEK in nature preservation and conservation. Although the approaches of science and TEK differ greatly, the combination of the two approaches strengthens both, as well as nature preservation as a whole. In many different parts of the world, indigenous cultures have developed through relationships with the land that are based on survival and respect. Indigenous peoples in nature-based cultures use their TEK to preserve their cultures and manage natural resources by protecting the land and the biodiversity with which they are familiar. Because the ecological knowledge of indigenous peoples is developed through experience, it can be preserved only if the cultures of the peoples are also preserved.

Around the world, about sixty million indigenous persons rely on forests for survival and for cultural identity. In the Amazon, indigenous peoples have a strong connection to and extensive knowledge of the rain forests. Three-quarters of the national parks in Latin America include land that is inhabited by indigenous peoples, and globally, up to 85 percent of protected areas are inhabited by indigenous peoples.

Most TEK is passed orally in cultures that have no written history, and scientists who recognize the value of TEK attempt to learn from indigenous peoples how they live in their natural environments, how they relate to the land, and how they effectively conserve the resources in their environments. An example of such efforts is a study that was conducted from 1995 to 1998 by scientists seeking to understand TEK on beluga whales. This research included interviews with members of indigenous communities in Alaska and Russia to learn from their experiences with and knowledge of beluga whales. The researchers found that through indirect, conversation-style interviews with indigenous persons, both individually and in groups, they were able to gather unexpected and insightful information that would not have been revealed through question-by-question interviews.

## Conflicts and Controversies

TEK is gradually becoming an accepted supplement to Western science and methods of nature preservation, but progress in accepting the importance of TEK is slow because of the many differences between TEK and mainstream scientific approaches to environmental issues. In contrast with the quantitative and mechanistic character of mainstream science, for example, TEK is qualitative and spiritual.

Because of differences in cultures, values, and approaches to knowledge, conflicts arise between the interests of indigenous peoples and those of Western science and the modern world. The indigenous peoples of the Amazon rain forests, for example, have interests that are in opposition to the modern world's commercial interests in the oil, minerals, and timber found in regions where indigenous communities live. In a world of commercially interested societies, indigenous cultures exist under great threat, as reports have suggested that indigenous peoples' lands hold the majority of the earth's remaining natural resources, including minerals, oil, fresh water, and medicinal plants.

Another conflict between indigenous peoples and

Western scientists concerns the commercial use of TEK. Indigenous peoples have profound ecological knowledge based on generations of firsthand experience, much of which includes natural remedies, access to unique plants, and TEK that could prove useful in bioscience. For example, Brazilian indigenous communities traditionally use a poison produced by a certain species of frog as a strong painkiller, and this substance has become part of numerous patent requests in both Europe and the United States. The exploitation by outsiders of the biodiverse resources of the lands where indigenous peoples live (referred to as bioprospecting or biopiracy, depending on the viewpoint of the speaker) is the source of ongoing controversy.

Many countries have made treaties with the indigenous peoples located within their borders, often regarding land rights and federal recognition. The United States officially recognizes approximately 560 Native American tribes, and hundreds of other tribes are in the process of trying to attain federal recognition. The status of U.S. federal tribal recognition establishes a government-to-government relationship in which the tribe is recognized as a sovereign nation. Federal recognition is beneficial for Native American tribes that are active in environmental issues and preservation because the tribes are treated as separate governments with their own lands to preserve or manage. In the United States, federal funding and other benefits aid tribes in establishing their own natural resources departments, which are necessary for nature preservation as well as for preserving the tribes' cultures.

## Climate Change

Climate change is a well-recognized environmental issue that affects indigenous peoples across the globe. Some indigenous populations live in societies with modern cultures, such as the indigenous tribes of the United States, but for many of the world's indigenous peoples climate change poses risks to their ancestral lands, to the sustainability of their food sources, and to the traditional ways of their cultures.

The potential effects of climate change on indigenous peoples' agricultural practices are particularly important because many indigenous cultures rely on farming for survival. Changes in climate conditions that result in droughts, floods, hurricanes, or frosts can have serious consequences for traditionally agricultural cultures, displacing populations and changing their relationship to the land.

Rising sea levels pose a climate change-related risk to island and coastal indigenous cultures, which may be directly affected through coastal land inundation that forces people to relocate and increases population density. Residents of the Alaskan village of Newtok, the majority of whom are Yupik Alaska Natives, began being driven from their homes in 2007 by the melting of the permafrost on which the village was built. Many other communities of Alaska Natives are likely to face similar situations in the future if climate change progresses as scientists have projected it will. Such displacements of indigenous populations put cultures as well as communities at risk.

In April, 2009, the Inuit Circumpolar Council hosted the Indigenous Peoples' Global Summit on Climate Change in Anchorage, Alaska, which focused on the effects of climate change on indigenous peoples. Among the topics discussed by the four hundred participants in the summit was the role that indigenous peoples might play in developing the successor agreement to the Kyoto Protocol, which was to be addressed at the United Nations Climate Change Conference held in Copenhagen, Denmark, in December, 2009. The participants also discussed the role that TEK could play in the mitigation of and adaptation to climate change.

*Allyson Leigh Hughes*

## Further Reading

Colchester, Marcus. "Self-Determination or Environmental Determinism for Indigenous Peoples in Tropical Forest Conservation." *Conservation Biology* 14, no. 5 (2000): 1365-1367.

Mauro, Francesco, and Preston D. Hardison. "Traditional Knowledge of Indigenous and Local Communities: International Debate and Policy Initiatives." *Ecological Applications* 10, no. 5 (2000): 1263-1269.

Menzies, Charles R., ed. *Traditional Ecological Knowledge and Natural Resource Management.* Lincoln: University of Nebraska Press, 2006.

Posey, Darrell Addison, and Michael J. Balick, eds. *Human Impacts on Amazonia: The Role of Traditional Ecological Knowledge in Conservation and Development.* New York: Columbia University Press, 2006.

Selin, Helaine, ed. *Nature Across Cultures: Views of Nature and the Environment in Non-Western Cultures.* Norwell, Mass.: Kluwer Academic, 2003.

Westra, Laura. *Environmental Justice and the Rights of Indigenous Peoples: International and Domestic Legal Perspectives.* Sterling, Va.: Earthscan, 2008.

# International Union for Conservation of Nature

CATEGORIES: Organizations and agencies; preservation and wilderness issues; animals and endangered species; resources and resource management

IDENTIFICATION: Environmental organization devoted to addressing issues of conservation and sustainable development

DATE: Founded on October 5, 1948

SIGNIFICANCE: The International Union for Conservation of Nature brings together governments, nongovernmental organizations, scientists, communities, and commercial interests in its efforts to protect species and ecosystems and promote sustainable use of natural resources. It has been instrumental in drafting and gaining signatories for a number of important international treaties designed to protect the environment.

The International Union for Conservation of Nature (IUCN) is the world's oldest global environmental organization and its largest professional global conservation network. Scientists working within IUCN gather data, identify species and wildlife areas in need of protection, and determine ways to sustain the earth's resources for present and future generations.

This democratic membership union, headquartered in Gland, Switzerland, serves as a neutral forum where governments, scientists, nongovernmental organizations, commercial interests, and communities come together to find solutions for conservation and development concerns. Its membership spans 140 countries and includes more than two hundred governmental organizations and more than eight hundred nongovernmental organizations. IUCN employs the voluntary expertise of almost eleven thousand scientists and specialists to provide policy advice on conservation issues. A leading authority on sustainability and the environment, the organization develops and supports conservation science and runs thousands of field projects and activities around the world.

The idea for an international conservation organization emerged as early as 1910. Swiss naturalist Paul Sarasin advocated for one and succeeded in establishing it, but it lost momentum during World War I. Sarasin was unable to reestablish a viable organization before his death in 1929, and none of the other persons who had been inspired by his vision were able to make significant headway over the next two decades. The establishment of the United Nations after World War II, however, meant new hope for Sarasin's dream. In 1948, an international conference on nature conservation was held at Fontainebleu, France, and a majority of the delegates in attendance—representatives from twenty-three governments, 126 national institutions, and eight international organizations—signed a formal act that established the International Union for the Protection of Nature (IUPN) on October 5.

Originally headquartered in Brussels, Belgium, IUPN pledged to protect endangered species of wildlife on a worldwide scale. It created an international network of conservationists who used the union and its periodic conferences as means for sharing information. In 1950 IUPN launched the Survival Service (later the Species Survival Commission), a group of scientists who worked on a volunteer basis to document the plights of the world's endangered species. IUPN began to interact directly with government representatives from many nations to promote conservation of their endangered species.

In 1956 IUPN changed its named to the International Union for Conservation of Nature and Natural Resources, or IUCN. Between 1990 and 2008, the name World Conservation Union was also used in conjunction with the IUCN name.

### THE ORGANIZATION MATURES

In 1961, IUCN moved its headquarters from Belgium to Switzerland. That same year it welcomed the fledgling World Wildlife Fund (WWF) to share its new headquarters. WWF was created to raise funds for IUCN and other worldwide conservation efforts, to conduct public relations campaigns, and to gain public support for conservation efforts. Ornithologist Peter Scott, then vice president of IUCN, became WWF's first chairperson. IUCN was among the earliest recipients of WWF grants, which funded projects such as creating a footpath in a reserved forest of Madagascar, transporting eight endangered white rhinos from South Africa to what is now Zimbabwe for breeding, and saving the rare Arabian oryx from extinction. The two organizations shared accommodations until the burgeoning WWF moved its offices in 1979. (WWF changed its name to the World Wide Fund for Nature in 1986 but retained use of the acronym by which it has been known since its founding.)

In 1962, the United Nations General Assembly

adopted the first World List of National Parks and Equivalent Reserves, a list more than three hundred pages long compiled by IUCN's newly formed Commission on National Parks and Protected Areas (later its World Commission on Protected Areas). The commission subsequently worked to develop detailed criteria for what constitutes a protected area. The U.N. List of Protected Areas, a listing of the world's national parks, scientific reserves, and natural monuments, later became a regular publication of IUCN.

In 1963 IUCN began its Red List system for identifying the world's threatened species and assessing conservation efforts. The Red List has since become recognized as the authoritative global listing of plants and animals facing possible extinction.

IUCN and the United Nations Educational, Scientific, and Cultural Organization (UNESCO) codrafted the text of the Convention Concerning the Protection of the World Cultural and Natural Heritage in 1972. This convention established the World Heritage List, which names valued natural and cultural sites to be protected. IUCN continues to serve as a technical advisory body to UNESCO's World Heritage Committee.

In 1975 the Convention on International Trade in Endangered Species of Wild Fauna and Flora (CITES) went into force. This international agreement, created to ensure that trade in specimens of wild animals and plants does not jeopardize the survival of species, emerged from a 1963 IUCN member resolution. IUCN teamed with WWF in 1976 to set up a monitoring network called the Trade Records Analysis of Flora and Fauna in Commerce (TRAFFIC). TRAFFIC investigators found that the smuggling and poaching of animal and plant species was occurring throughout the world, even in countries that had signed treaties opposing such activities. Increased awareness of the problem encouraged local governments to work with TRAFFIC to stop illegal commerce in exotic animals and plants.

In 1980 IUCN, WWF, and the United Nations Environment Programme issued a joint publication titled *World Conservation Strategy: Living Resource Conservation for Sustainable Development*, which emphasized the need for a holistic approach to conservation and the sustainable use of natural resources. Fifty countries soon created their own national conservation strategies based on the recommendations presented in *World Conservation Strategy*. The three organizations copublished a follow-up strategy document, *Caring for the Earth: A Strategy for Sustainable Living*, in 1993, in which they recommended 132 actions that individuals at all social and political levels can take to protect the environment and improve quality of life.

Other major IUCN contributions include preparation of the World Charter for Nature (1982), the Convention on Biological Diversity (1992), and the draft Covenant on Environment and Development (1995; updated 2000, 2004, and 2010). IUCN also sponsors the World Conservation Congress, which convenes every four years.

*Lisa A. Wroble*
*Updated by Karen N. Kähler*

FURTHER READING

Bräutigam, Amie, and Martin Jenkins. *The Red Book: The Extinction Crisis Face to Face*. Gland, Switzerland: IUCN, 2001.

Holdgate, Martin W. *From Care to Action: Making a Sustainable World*. Washington, D.C.: Taylor & Francis, 1996.

International Union for Conservation of Nature, United Nations Environment Programme, and World Wide Fund for Nature. *Caring for the Earth: A Strategy for Sustainable Living*. Gland, Switzerland: Authors, 1993.

_____. *World Conservation Strategy: Living Resource Conservation for Sustainable Development*. Gland, Switzerland: Authors, 1980.

Morphet, Sally. "NGOs and the Environment." In *The Conscience of the World: The Influence of Non-governmental Organizations in the U.N. System*, edited by Pete Willetts. Washington, D.C.: Brookings Institution Press, 1996.

Van Dyke, Fred. *Conservation Biology: Foundations, Concepts, Applications*. 2d ed. New York: Springer, 2008.

Vié, Jean-Christophe, Craig Hilton-Taylor, and Simon N. Stuart, eds. *Wildlife in a Changing World: An Analysis of the 2008 IUCN Red List of Threatened Species*. Gland, Switzerland: IUCN, 2009.

# John Muir Trail

CATEGORY: Preservation and wilderness issues
IDENTIFICATION: Wilderness path through the high country of California's Sierra Nevada range
SIGNIFICANCE: A unique hiking footpath, the John Muir Trail passes through what many backpackers

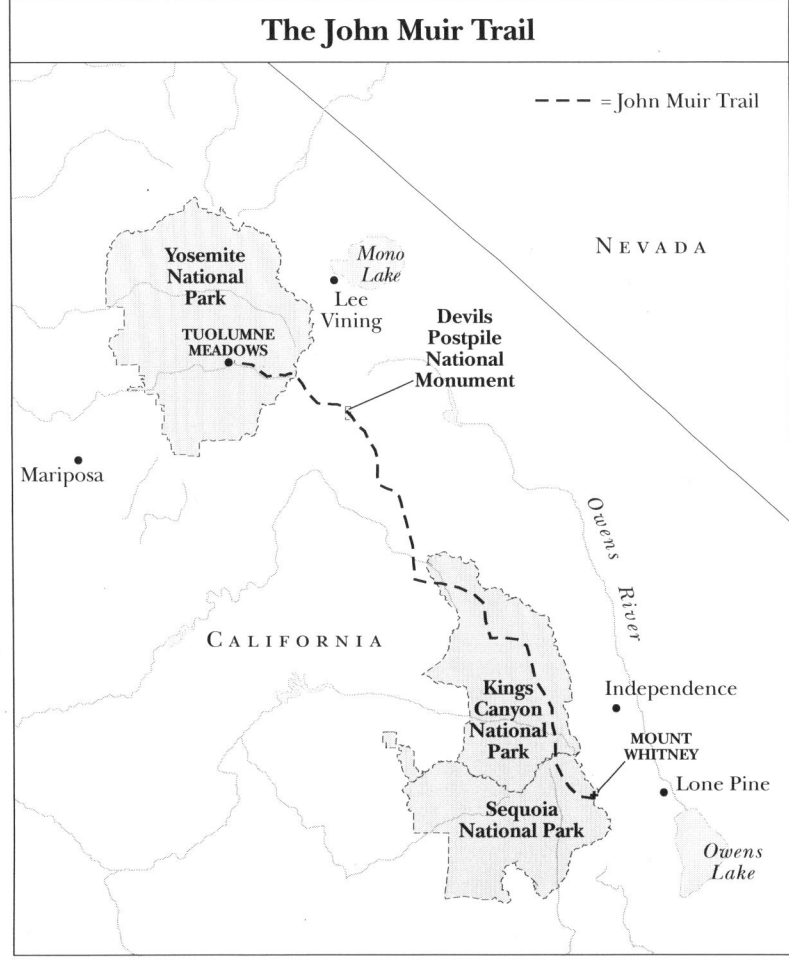

### The John Muir Trail

regard as the finest mountain scenery in the United States and among the finest in the world. Its scenery includes spectacular rugged granite cliffs, lakes, streams, waterfalls, diverse wildlife, and abundant wildflowers.

The John Muir Trail runs south 340 kilometers (211 miles) from Yosemite Valley to the summit of Mount Whitney, the highest mountain in the continental United States. Without crossing a single road, the trail goes through Yosemite, Kings Canyon, and Sequoia national parks, as well as national forest land and Devils Postpile National Monument. After rising from the floor of Yosemite Valley it rarely drops below an altitude of 2,400 meters (8,000 feet) and crosses six passes at altitudes above 3,400 meters (11,000 feet), the highest of which is Forester Pass at 4,009 meters (13,153 feet).

The mountain region through which the John Muir Trail passes was not explored until the late 1850's and the 1860's. Among the first to get deep into the backcountry was the Scottish American naturalist and environmentalist John Muir. A geologist, Josiah Whitney, was tasked with making a geological survey of the state of California in 1864; that summer he led a team of scientists in exploring and surveying the area through which the trail would eventually wind. In 1884 a fourteen-year-old boy, Theodore Solomons, was the first to come up with the idea of a crest-parallel trail from Yosemite to Kings Canyon. Along with Solomons, during the 1880's and 1890's Joseph N. "Little Joe" LeConte and Bolton Brown explored routes and passes in the area.

In 1914 a member of the Sierra Club, after hiking the area, proposed that the club ask the California State Legislature to fund a high-mountain trail. The club took up the cause and gained the legislature's support for funding, and construction started in 1915, one year after John Muir died. The building of the trail was a very difficult task, as most of the country it would go through was extremely rugged, much of it consisting of very steep granite slopes. In addition, all of the materials and tools used had to be carried or packed in. The trail was completed in 1938.

*C. Mervyn Rasmussen*

## Kings Canyon and Sequoia national parks

CATEGORIES: Places; preservation and wilderness issues
IDENTIFICATION: U.S. national parks in California's southern Sierra Nevada range
DATES: Established in 1890 (Sequoia) and 1940 (Kings Canyon)

SIGNIFICANCE: An ongoing environmental challenge can be seen in the struggle to maintain a balance between preserving the environment in Kings Canyon and Sequoia national parks and allowing the public to have access to the parks' lands for recreational purposes.

Sequoia National Park and Grant Grove were established in 1890 in California's southern Sierra Mountains. In 1940 Grant Grove was incorporated into the new Kings Canyon National Park. The impetus for the creation of both Kings Canyon and Sequoia came from the work of such preservationists as John Muir and the Sierra Club, as well as from the desires of local inhabitants of the San Joaquin Valley. When the National Park Service was created in 1916, it established the obligation of all national parks in the United States to provide public resources for recreation as well as to preserve the natural environment. These two aims have proved to be difficult to reconcile, however, particularly in Sequoia and Kings Canyon.

By the end of the 1920's it was apparent to some that environmental damage was already occurring in the Giant Forest area of Sequoia. Even before 1890 people had enjoyed camping among the redwoods, and it was inevitable that concessionaires would construct cabins and other buildings in the shadows of the great trees. Their fragile root systems were negatively affected by the buildings, the sewer system, and the human traffic that grew over time. It was not until the 1990's, however, that the buildings were actually torn down, to be relocated in areas where they would not have detrimental impacts on the redwoods.

*A hiker is dwarfed by giant redwood trees in Sequoia National Park. (©Jim Lopes/iStockphoto.com)*

Fire prevention is another concern of national park officials. Time and knowledge have radically altered the approaches taken toward fire prevention in Sequoia, as well as in other national parks. In the 1960's park officials realized that periodic fires are necessary to maintain the health of forests because they remove dead material and allow new seedlings to sprout and flourish. A living forest requires fire, and controlled burns became common in the park.

Although most visitors to Sequoia and Kings Canyon get no farther than the Giant Forest and General Grant areas, both similar in their redwood ambience, both parks are mainly wilderness areas of streams, meadows, and some of the nation's most rugged mountains. Before the parks were established, the areas had been overgrazed by sheep and cattle. Although this problem ended when the parks were founded, overuse by campers and backpackers continued to be a challenge. Eventually, access to the backcountry was limited through a permit system, the overused meadows were temporarily closed, and all visitors were required to pack out everything that they carried in to minimize damage to the fragile wilderness environment.

Kings Canyon has faced environmental challenges. Its dramatic central valley along the Kings River has been subject to numerous requests for dam construction, even within the park, and requests that the valley floor be developed for additional visitors. Here, too, preservation has taken precedence over recreation and other development, and human-built facilities have been limited intentionally.

Preservation of the environment and of ecosystems has continued to take precedence over mere recreation in both parks, with the public generally understanding and supporting the Park Service's priorities. However, the parks have been subject to threats by forces outside the Park Service's control, particularly air pollution wafting up from the floor of the valley below and, further afield, the possible long-term dangers of climate change.

*Eugene Larson*

FURTHER READING

Beesley, David. *Crow's Range: An Environmental History of the Sierra Nevada.* Reno: University of Nevada Press, 2004.

Eldredge, Ward. *Kings Canyon National Park.* Charleston, S.C.: Arcadia, 2008.

Noss, Reed F., ed. *The Redwood Forest: History, Ecology, and Conservation of the Coast Redwoods.* Washington, D.C.: Island Press, 2000.

## Land-use planning

CATEGORIES: Land and land use; urban environments

DEFINITION: Systematic development of a vision and public policy that guides the future use of land, facilities, and resources

SIGNIFICANCE: Land-use planning affects the livability and sustainability of a community by establishing social, economic, and environmental policies. Land-use plans provide a basis for the efficient use of land, for community improvements and economically viable development, and for the preservation and management of the environment and existing resources.

Governments engage in land-use planning for many reasons. Land-use plans support the adoption of legitimate land-use regulations that will withstand scrutiny in a court of law. Land-use planning results in proposals and policies that guide future land development and the installation of infrastructure to support such uses. Infrastructure planning requires the projection of the locations of roads and transportation networks, sewers, water lines, and other utilities and the consideration of future budgeting to allow for the construction of such improvements. Moreover, as development that is based on a land-use plan becomes a reality, a community is able to budget in advance for capital improvement facilities such as schools, fire stations, and municipal buildings. By adopting land-use plans, cities and towns are able to achieve sustainable development through the integration of existing and future land uses and natural systems with existing and future infrastructure and major projects. Ideally, land-use planning promotes order and economic change while preserving the environment.

A major specialty in land-use planning is urban planning, which considers the built environment and attempts to address some kinds of social problems, such as by finding ways to discourage crime. Land-use planning for urban environments has existed since ancient times, when the first human settlements were designed and built to achieve order, efficiency, aesthetics, and safety and to meet the functional needs of society, including community services, water, sanitation, transportation, and military defense. Modern urban land-use planning may involve urban renewal proposals that encourage new investment and development in decaying and declining areas of a city.

SUSTAINABLE DEVELOPMENT

The United Nations has been in the forefront of the sustainable development movement, which involves the efficient marriage of economic development, environmental protection, and sociopolitical policy. It is quite challenging to respect the carrying capacity of the natural environment while at the same time encouraging economic development, and land-use planning can be a viable tool for achieving sustainable development that will meet both the present and the future needs of a society.

To contribute to the achievement of sustainable development, those involved in land-use planning must consider whether the natural environment will be able to withstand certain social policies, such as bringing jobs to a community through industrial development, and whether certain environmental policies, such as the adoption of stringent environmental

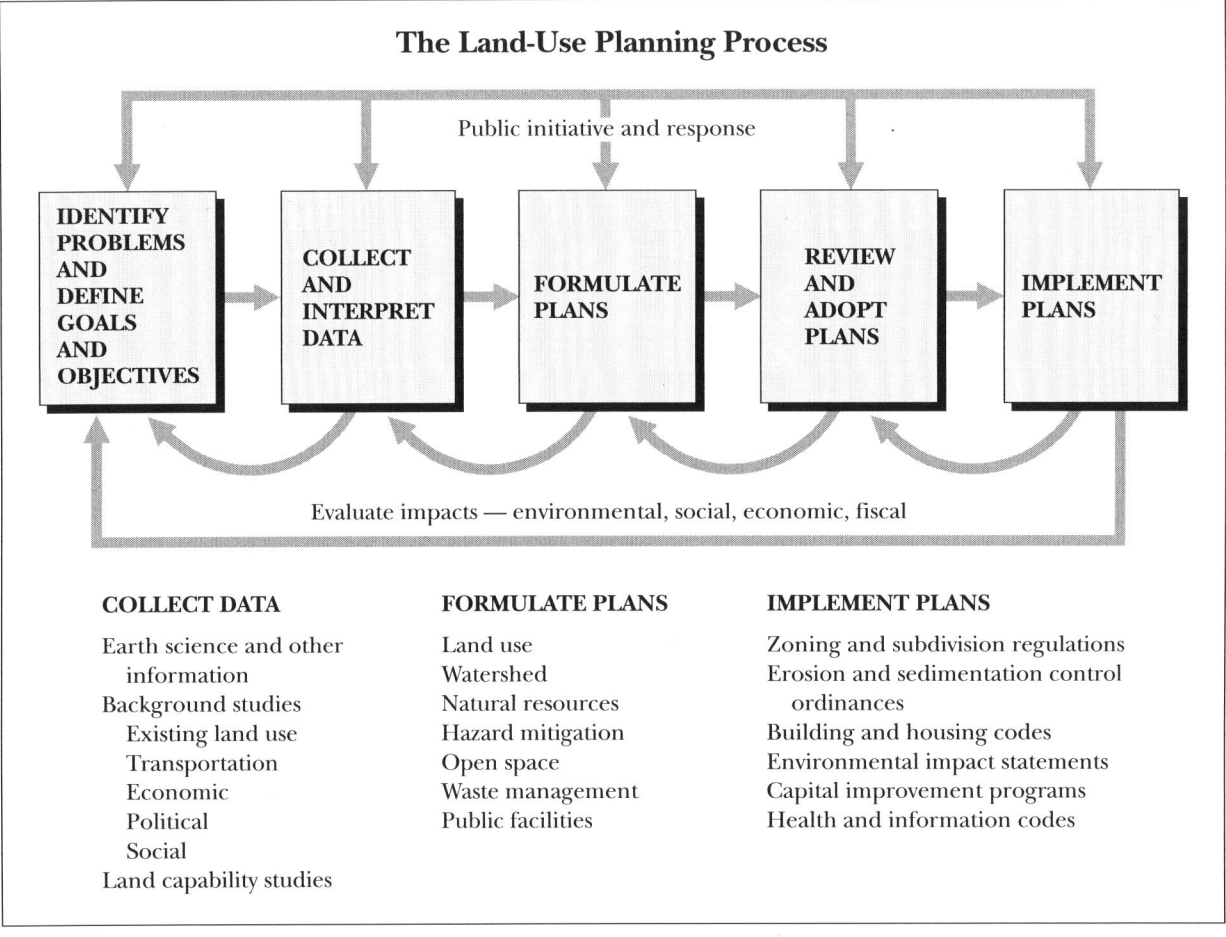

regulations, will thwart economic development. In addition, economic development must not deplete natural resources faster than they can be replenished, and sociopolitical policies should achieve equitable development that does not benefit one segment of society at the expense of another.

THE PLANNING PROCESS

Land-use planning is multidisciplinary and brings together many divergent views of practitioners such as architects, landscape architects, regional planners, urban designers, engineers, transportation planners, and environmentalists. Before a land-use plan becomes final, planning experts conduct numerous background studies and analyses. Examples include analyses of the suitability and carrying capacity of land for various land uses. A suitability study might consider not only the soils and topography of an area but also the availability of supportive infrastructure for future development. Another goal of land-use planning may be to conserve land and protect valuable ecosystems while allowing the public to make some use of these community assets. Planners may thus be concerned with environmental planning and might conduct ecological assessments of flora and fauna habitats. Planners may also consider watershed and groundwater management while concomitantly proposing viable locations for water-related infrastructure and facilities to support future development. Other important components of a land-use plan concern the management of sewers, stormwater, and solid waste.

Planners develop land-use plans and policies and present them to legislative bodies for adoption. As a plan evolves, land-use planners may take advantage of new technologies, such as digital media and geospatial data and information systems, that enable them to gain a better understanding of the interplay between the area's natural and developed land systems and the environmental science necessary to mitigate the impacts of past and existing land-use practices. Gaining approval for land-use plans requires planners to engage in negotiation with both government officials and interested members of the public to arrive at acceptable proposals supported by all parties. Land-use plans are the basis for many land-use regulations, including zoning, subdivision control, building codes, and environmental regulations. Often governments must rely on land-use plans to overcome legal challenges concerning the constitutionality of development regulations.

*Carol A. Rolf*

FURTHER READING

Berke, Philip R., and David R. Godschalk. *Urban Land Use Planning*. 5th ed. Champaign: University of Illinois Press, 2006.

Haar, Charles M., and Michael A. Wolf. *Land Use Planning and the Environment*. Washington, D.C.: Environmental Law Institute, 2009.

Juergensmeyer, Julian Conrad, and Thomas E. Roberts. *Land Use Planning and Development Regulation Law.* 2d ed. Eagan, Minn.: Thomson West, 2006.

Randolph, John. *Environmental Land Use Planning and Management*. Washington, D.C.: Island Press, 2003.

Silberstein, Jane, and Chris Maser. *Land-Use Planning for Sustainable Development*. Boca Raton, Fla.: CRC Press, 2000.

# Leopold, Aldo

CATEGORIES: Activism and advocacy; preservation and wilderness issues
IDENTIFICATION: American wilderness conservationist and environmental philosopher
BORN: January 11, 1887; Burlington, Iowa
DIED: April 21, 1948; near Baraboo, Sauk County, Wisconsin
SIGNIFICANCE: Leopold, who has been called the father of modern wildlife management and ecology, applied his insightful concepts of ethics and philosophy to conservation strategies and thus helped raise awareness of environmental issues.

Aldo Leopold developed an interest in wildlife while observing waterfowl and animals living in the Mississippi River marshes near his childhood home. After completing graduate studies at the Yale School of Forestry in 1909, he worked for the U.S. Forest Service in the Arizona Territory and New Mexico. With his colleague Arthur Carhart, Leopold demanded that the Forest Service preserve wilderness areas in national forests for their aesthetic and recreational value. As a result, in June, 1924, the Forest Service set aside 305,500 hectares (755,000 acres) in New Mexico as the Gila Wilderness Area, initiating the protection of millions of forest acres for environmentalists to study and enjoy.

Leopold then initiated a game protection movement in the southwestern United States. An avid sportsman, he initially promoted the hunting of predators to protect game. When a deer herd on the Arizona Kaibab Plateau suffered starvation because predator-control measures, among other factors, had resulted in a great increase in the deer population, Leopold realized that the predator-prey balance is crucial to a healthy, stable environment. As game consultant for the Sporting Arms and Ammunition Manufacturers' Institute, he published one of the first studies of American game populations, titled *Report on Game Survey of the North Central States* (1931). He traveled to Europe to assess the game management techniques used in various countries, and he developed a national game management policy for the American Game Protective Association.

In 1933 Leopold published *Game Management*, a popular textbook in which he discussed wildlife population dynamics and habitat protection, emphasizing the preservation of ecosystems. He considered the wilderness as a community to be shared rather than a commodity to be controlled by humans, stressing that nature should be treated ethically and not appropriated for economic gain. Leopold accepted the chair in game management, created especially for him, at the University of Wisconsin in 1933. He built a cabin by the Wisconsin River, planted trees, experimented with land restoration, and observed the wildlife of Sauk County. Active in conservation groups, he helped organize the Wilderness Society and the Wildlife Society. President Franklin D. Roosevelt appointed Leopold to the Special Committee on Wild Life Restoration in 1934.

Leopold wrote a number of essays about environmental ethics, which were compiled in *A Sand County Almanac, and Sketches Here and There* (1949) after his death. He outlined his concept of a land ethic in which all wilderness residents have a vital role. He believed that humans are a part of nature and that all persons have a responsibility to understand, protect, and live in harmony with a healthy environment. He argued that humans should respect nature and "preserve the integrity, stability, and beauty of the biotic community." Because of its literary appeal, Leopold's book introduced a broader audience to the conservation movement. When the book was reissued in the 1960's, a new generation of environmental readers embraced Leopold's passion for nature, praising his conservation initiatives and wilderness wisdom.

*Elizabeth D. Schafer*

FURTHER READING

Knight, Richard L., and Suzanne Riedel, eds. *Aldo Leopold and the Ecological Conscience*. New York: Oxford University Press, 2002.

Leopold, Aldo. *A Sand County Almanac, and Sketches Here and There*. 1949. Reprint. New York: Oxford University Press, 1987.

Lorbiecki, Marybeth. *Aldo Leopold: A Fierce Green Fire*. 1996. Reprint. Guilford, Conn.: Globe Pequot Press, 2005.

## Logging and clear-cutting

CATEGORY: Forests and plants

DEFINITIONS: Logging is the harvesting of timber from forestlands with the intention of using it for specific purposes, such as lumber, fuelwood, or the production of pulp or chemicals; clear-cutting is a logging technique in which all the timber is removed from a stand at the same time

SIGNIFICANCE: Improper logging and clear-cutting can pose a variety of threats to the environment, including the destruction of wildlife habitat and the disturbance of soil in ways that can lead to erosion and flooding. Environmentally sensitive logging practices can minimize such negative impacts.

Although some people may think of logging and clear-cutting as practically synonymous, the two are not the same. Similarly, many people may have the impression that commercial logging is responsible for all losses of forestland, but, particularly in tropical areas, many hectares of forestland are cleared annually for other purposes. Rain forests in Amazonia, for example, are often bulldozed to create pastureland for cattle. The timber is not harvested; rather, it is simply pushed into piles and burned at the site.

Logging and clear-cutting, if improperly done or motivated by short-term economic goals, can pose significant threats to the environment. Logging always involves some disturbance to soil and wildlife. If performed in environmentally sensitive areas, it can destroy irreplaceable habitat, contribute to problems with erosion and flooding, and worsen the threat of global warming. Heavy equipment can compact soil, leaving ruts that may persist for many years, while clear-cutting hillsides can lead to erosion, stream siltation, and devastating floods. In Asia, for example, clear-cutting in the mountains of Nepal and India resulted in disastrous floods in Bangladesh. Even when logging does not inflict long-term damage on the immediate environment, the simple removal of trees can contribute to global warming. The burning of slash (waste material) at logging sites pumps greenhouse gases into the atmosphere, and the loss of forest means that there are fewer trees to break those gases down into oxygen and organic compounds.

Regardless of whether a logger is cutting only one tree or one thousand, logging involves four basic steps: selecting the timber to be harvested, felling the trees, trimming away waste material, and removing

## U.S. Lumber Consumption
(billions of board feet)

| | 1995 | 2000 | 2003 | 2005 | 2007 |
|---|---|---|---|---|---|
| **Species group** | | | | | |
| Softwoods | 47.6 | 54.0 | 56.5 | 64.4 | 50.5 |
| Hardwoods | 11.7 | 12.2 | 10.5 | 11.2 | 10.3 |
| **End use** | | | | | |
| New housing | 15.9 | 20.6 | 24.0 | 28.6 | 26.2 |
| Residential | 14.3 | 16.4 | 18.3 | 20.6 | 20.9 |
| New nonresidential construction | 5.8 | 5.1 | 4.4 | 4.3 | 3.9 |
| Manufacturing | 5.5 | — | 8.1 | 7.7 | 7.3 |
| Shipping | 8.5 | 7.4 | 7.5 | 7.6 | 7.7 |
| Other | 9.3 | 16.1 | 4.7 | 7.0 | — |

*Source:* U.S. Forest Service, *U.S. Timber Production, Trade, Consumption, and Price Statistics, 1965-1999*, 2001; and U.S. Department of Commerce, *Statistical Abstract of the United States, 2009*, 2009.

*A clear-cut mountainside in Canada.* (©Charles Dyer/Dreamstime.com)

the desired portion of the tree from the woods. Equipment used in logging ranges from simple hand tools, such as axes and crosscut saws, to multifunction harvesting machines costing hundreds of thousands of dollars each. A mechanized feller buncher, for example, can fell a tree, trim off the branches, cut the stem into logs of the desired length, and stack the logs to await removal from the forest. The choice of equipment utilized in harvesting any specific stand of timber depends on factors such as the terrain, the type of timber to be logged, and whether the logger intends to harvest only selected trees or to clear-cut the site.

Loggers are more likely to clear-cut, or remove all the standing timber from a section of land, if the timber is plantation grown and of a uniform age and size. Clear-cutting also occurs in forests where the desired species of trees need large amounts of sunlight to regenerate. Many conifers, such as Douglas fir, are shade-intolerant. Landowners will occasionally decide to change the dominant species on a tract and so will clear-cut existing timber to allow for replanting with new, more commercially desirable trees. Clear-cutting can be an acceptable practice in sustainable forestry when plantation stands are harvested in rotation.

Selective harvesting, in contrast with clear-cutting, leaves trees standing on the tract. Selective harvesting can be utilized with even-age plantation stands as a thinning technique. More commonly, it is used in mixed and uneven-age stands to harvest only trees of desired species or sizes. In cutting hardwood for use as lumber, for example, 30 centimeters (12 inches) may be considered the minimum diameter of a harvestable tree. Trees smaller than that will be left in the woods to continue growing.

An individual, noncommercial woodcutter may fell only a few trees per year on small parcels of land. Commercial loggers, in contrast, annually harvest hundreds of thousands of trees and operate on large parcels of land. Nonetheless, a significant number of hectares of forestland are cleared annually by people who rely on noncommercial logging for wood for their own individual needs, such as fuel for cooking or heating their homes. Although some woodcutters may cut more than they need for their own use and then sell the surplus, fuelwood for individual households is usually gathered by members of the households that use the wood. Other examples of noncommercial logging include farmers cutting trees for use as fencing or building materials on their own property.

From an environmental viewpoint, the biggest difference between commercial and noncommercial logging would seem to be one of scale, but even this is not always true. Improper logging on a small parcel can have more of an impact on a watershed or ecosystem than professional harvesting of a large stand. Even if no single household's logging practices pose a problem, the gathering of fuelwood or other timber by multiple households in an area can be devastating. With no guidance from professional foresters, woodcutters tend to harvest trees based on convenience for themselves rather than on principles of sustainable forestry or watershed management. Many nations have developed programs in which professional foresters provide advice to small property owners on environmentally sound timber harvesting practices and improvement of timber stands, but the availability of such help varies widely from country to country.

*Nancy Farm Männikkö*

### Further Reading

Bevis, William W. *Borneo Log: The Struggle for Sarawak's Forests*. Seattle: University of Washington Press, 1995.

Colfer, Carol J. Pierce. *The Equitable Forest: Diversity, Community, and Resource Management*. Washington, D.C.: Resources for the Future, 2005.

Davis, Lawrence S., et al. *Forest Management: To Sustain Ecological, Economic, and Social Values*. 4th ed. Boston: McGraw-Hill, 2001.

Stenzel, George, Thomas A. Walbridge, Jr., and J. Kenneth Pearce. *Logging and Pulpwood Production*. 2d ed. New York: John Wiley & Sons, 1985.

Tacconi, Luca, ed. *Illegal Logging: Law Enforcement, Livelihoods, and the Timber Trade*. Sterling, Va.: Earthscan, 2007.

Walker, Laurence C., and Brian P. Oswald. *The Southern Forest: Geography, Ecology, and Silviculture*. Boca Raton, Fla.: CRC Press, 2000.

## Lovejoy, Thomas E.

CATEGORIES: Activism and advocacy; preservation and wilderness issues
IDENTIFICATION: American tropical biologist
BORN: August 21, 1941; New York, New York
SIGNIFICANCE: Lovejoy is recognized for his contributions to conservation policy making. He is best known for developing creative solutions to issues of scientific concern, such as debt-for-nature swaps.

In 1971 Thomas E. Lovejoy earned a doctor of philosophy degree in biology from Yale University. His early research efforts included a long-range study of birds in the Brazilian Amazon. From 1973 to 1987 Lovejoy served as program director for the World Wildlife Fund (WWF) and was responsible for the organization's projects in the Western Hemisphere and those involving tropical forests. He is credited with being the first person to use the term "biological diversity," or "biodiversity," in 1980. Lovejoy also began speaking of global extinction rates, and in 1980 his first projection of such rates appeared in *The Global 2000 Report*, submitted to U.S. president Jimmy Carter. This marked the beginning of his many contributions to the policy arena.

It was during his tenure as executive vice president of WWF from 1985 to 1987 that Lovejoy originated the concept of the debt-for-nature swap, which involves forgiveness of foreign debts incurred by developing nations in exchange for those nations' agreements to protect the fragile ecosystems within their borders. Debt-for-nature swaps have been used to preserve wilderness in countries such as Costa Rica, Bolivia, Ecuador, the Philippines, and Madagascar. While working for WWF, Lovejoy increased public interest in the issue of preserving tropical forests.

Lovejoy is recognized for his practical approach to solving environmental problems. This is illustrated by his input for the Minimum Critical Size of Ecosystems Project, also known as the Biological Dynamics of Forest Fragments Project. This was a joint research project undertaken by the Smithsonian Institution and Brazil's National Institute of Amazonian Research. The project was one of the earliest attempts to define minimum sizes for national parks and biological reserves and to formulate management strategies for such protected areas. In recognition of his conservation work in Brazil, Lovejoy became the first environmentalist to receive the Order of Rio Branco, which the Brazilian government awarded him in 1988.

In 1987 Lovejoy created the Public Broadcasting Service television series *Nature*, for which he served as adviser for many years. In the same year he was appointed assistant secretary for environmental and external affairs at the Smithsonian Institution. By 1994 Lovejoy had risen to become counselor to the secretary for biodiversity and environmental affairs at the Smithsonian. While still associated with the Smithsonian Institution, he became chief biodiversity adviser to the World Bank in 1998.

While continuing his association with the Smithsonian, Lovejoy also served in a variety of advisory positions. In 1993 he was chosen to be science adviser by the U.S. secretary of the interior and helped set up the National Biological Survey. From 1994 to 1997 Lovejoy served as scientific adviser to the executive director of the United Nations Environment Programme (UNEP), and he later became senior adviser to the president of the United Nations Foundation. His professional contributions have included serving as chair of the Yale Institute for Biospheric Studies, as chair of the U.S. Man and the Biosphere Program, as president of the Society for Conservation Biology, and as president of the American Institute of Biological Sciences. Among the honors Lovejoy has received are the 2001 Tyler Prize for Environmental Achievement and the 2002 Lindbergh Award, which is presented by the Lindbergh Foundation in recognition of contributions made to maintaining a balance between technology and preservation of the environment.

Lovejoy has published numerous articles and several books on environmental conservation, including two coedited volumes: *Global Warming and Biological Diversity* (1992; with Robert L. Peters) and *Climate Change and Biodiversity* (2005; with Lee Hannah).

Michele Zebich-Knos

### Further Reading

Lovejoy, Thomas E. "Biological Diversity." In *Life Stories: World-Renowned Scientists Reflect on Their Lives and on the Future of Life on Earth*, edited by Heather Newbold. Berkeley: University of California Press, 2000.

_____. "Climate Change and Prospects for Sustainability." In *Foundations of Environmental Sustainability: The Coevolution of Science and Policy*, edited by Larry L. Rockwood, Ronald E. Stewart, and Thomas Dietz. New York: Oxford University Press, 2008.

Lovejoy, Thomas E., and Lee Hannah, eds. *Climate Change and Biodiversity*. New Haven, Conn.: Yale University Press, 2005.

## Mather, Stephen T.

CATEGORIES: Activism and advocacy; preservation and wilderness issues
IDENTIFICATION: American conservationist
BORN: July 4, 1867; San Francisco, California
DIED: January 22, 1930; Brookline, Massachusetts
SIGNIFICANCE: As the first director of the U.S. National Park Service, Mather personified the national parks movement during the early decades of the twentieth century.

A descendant of one of America's early Puritan families, Stephen T. Mather was born and raised in California, where he came to love the mountains and forests of the western United States. He joined John Muir's Sierra Club in 1904. He was educated at the University of California at Berkeley, where he made many lifelong contacts. After graduation he went into the borax business, in which he accumulated considerable wealth.

Mather, a Republican, followed Theodore Roosevelt, the most famous conservationist in the United States, into the Progressive Party in 1912. However, Mather had the ability to transcend partisan politics. Several earlier attempts had been made to bring order to the small number of national parks that had already been established, and in 1913 President Woodrow Wilson's secretary of the interior, Franklin K. Lane, decided to upgrade the parks by appointing Adolph C. Miller, a fellow University of California graduate, as assistant secretary. After Miller was reassigned elsewhere, Lane challenged Mather to take over the position. He did so in 1915, with another Berkeley graduate, Horace Albright, as his chief assistant. In 1916 the Park Service Act became law, and Mather became the National Park Service's first and arguably most significant director.

As director of the service, Mather faced a series of interrelated issues: The parks required both financial and political support from Congress, which in turn depended on public opinion. Energetic, charismatic,

*Stephen T. Mather, right, with U.S. secretary of the interior Albert B. Fall at Glacier Point, Yosemite Valley, in 1921. (National Park Service Historic Photograph Collection)*

and brilliant at public relations, Mather generated popular support for the national park system through various avenues, including national publications such as the *Saturday Evening Post* and *National Geographic Magazine*—more than one thousand articles about the parks appeared from 1917 to 1919. He organized conferences, brought politicians and writers to the parks, and used his own funds in support of the park system; he personally paid the salary of Robert Sterling Yard, who became the parks' first publicist. In order to provide better access to the parks, Mather strongly supported at least some development; he worked with railroads and automobile associations in building roads and providing other public amenities in some parks. Through the years he had his differences with politicians, bureaucrats, and others, but he worked successfully with both Republicans and Democrats.

Mather retired in 1929 because of ill health and was succeeded by Albright, but he left the National Park Service as a major icon in America's consciousness. Under his tenure the park system increased from thirteen parks, eighteen national monuments, a total of 1.9 million hectares (4.8 million acres) of area, and 334,000 annual visitors to twenty national parks, thirty-two national monuments, 3.3 million hectares (8.3 million acres), and three million visitors each year. However, development engendered criticism, and Mather was accused of being too prodevelopment. Despite the criticism, he had no doubt about his responsibility as director; he stated, "Our job in the Park Service is to keep the national parks as close to what God made them as possible." No director since Mather's time has had as much impact on the U.S. National Park Service, and no one since has been able to generate as much public and political support for the park system.

*Eugene Larson*

FURTHER READING

Gonzalez, George A. "The Political Economy of the National Park System." In *Corporate Power and the Environment: The Political Economy of U.S. Environmental Policy.* Lanham, Md.: Rowman & Littlefield, 2001.

Heacox, Kim. *An American Idea: The Making of the National Parks.* Washington, D.C.: National Geographic Society, 2001.

Nash, Roderick. *Wilderness and the American Mind.* 4th ed. New Haven, Conn.: Yale University Press, 2001.

## Migratory Bird Act

CATEGORIES: Treaties, laws, and court cases; animals and endangered species
THE LAW: U.S. federal legislation designed to protect migratory birds
DATE: Became effective on March 4, 1913
SIGNIFICANCE: The Weeks-McLean Act was an important first step in U.S. government protection of migratory birds.

The Migratory Bird Act of 1913, also known as the Weeks-McLean Act (for its sponsors, Congressman John W. Weeks of Massachusetts and Senator John P. McLean of Connecticut), was preceded by the Lacey Act of 1900, which was introduced by John

Lacey of Iowa in the House of Representatives. Both laws were passed in an effort to protect wildlife, to eliminate the slaughter of wild birds and the interstate market for their parts for the lucrative millinery trade, and to establish a regulatory system of federal, state, and foreign statutes to govern and protect fowl and other wildlife.

It is estimated that during the late nineteenth century nearly 200 million birds were slaughtered for their feathers and carcasses, which were used to decorate women's hats. The calamitous species extinction of the passenger pigeon aroused the attention of conservationists and protectionists nationwide. George Bird Grinnell founded the first Audubon Society in 1886 in an effort to organize opposition to the bird trade. Ornithology and bird-watching were popular activities pursued by professionals and amateurs alike at the time. In 1896 Harriet Hemenway founded the Massachusetts Audubon Society and quickly attracted influential members who championed opposition to the wanton destruction of birds. President Theodore Roosevelt was an avid supporter of the Audubon Society. In 1910 New York State passed the Audubon Plumage Law, banning the sale of the feathers of birds native to the state.

In 1913 the Weeks-McLean Act put birds under federal jurisdiction, regulating the spring hunting season. This act was followed in 1916 by a convention signed by Canada and the United States to establish federal guidelines protecting both game and nongame birds on both sides of the border. Signed into law in the United States as the Migratory Bird Treaty Act of 1918, this treaty was amended in 1937 to include Mexico in a comprehensive plan to protect all migratory birds on the North American continent. The Weeks-McLean Act is considered one of the first environmental laws of the United States.

*Victoria M. Breting-García*

### Further Reading

Bean, Michael J. "Historical Background to the Endangered Species Act." In *Endangered Species Act: Law, Policies, and Perspectives*, edited by Donald C. Bauer and William Robert Irvin. 2d ed. Chicago: American Bar Association, 2010.

Burnett, J. Alexander. *A Passion for Wildlife: The History of the Canadian Wildlife Service*. Vancouver: University of British Columbia Press, 2003.

Price, Jennifer. *Flight Maps: Adventures with Nature in Modern America*. New York: Basic Books, 1999.

## Mine reclamation

CATEGORY: Land and land use
DEFINITION: Process of returning disturbed land areas to stable and productive uses after minerals have been removed through mining
SIGNIFICANCE: The reclamation of lands no longer used for mining is important to prevent various environmental problems, such as soil erosion, pollution of groundwater by acid drainage, and the physical dangers posed by abandoned shafts and tunnels.

Failure to reclaim mined land may result in substantial loss of biological productivity of the land surface, significant degradation of nearby water bodies, and hazards to human health and safety. Degradation of land and water is attributable to erosion and sedimentation of soils, acid mine drainage, and damage to groundwater aquifers. Hazards to human health and safety include open mine shafts, mine fires, subsidence of the ground above underground tunnels, clifflike surfaces known as highwalls, and landslides on steep slopes. Land areas disturbed by mining that have not been reclaimed are referred to as abandoned mine lands.

The amount of land to be reclaimed at any mine is determined by the amount of mineral removed and the type of mining operation. Underground mines require little reclamation except near the tunnel entrance. Surface or strip mines disturb larger areas and volumes of soil and rock than do underground mines, and therefore they require more reclamation. In all forms of surface mining, rock and soil located above and between seams of the mineral are removed to expose the mineral for extraction.

Before soil and rock can be removed, all vegetation covering the land surface to be mined must be removed. Next, topsoil and subsoils are excavated and used in an adjacent area that is being reclaimed, or they are separated from other overburden rock (rock that overlies deposits of minerals to be mined) and stockpiled for later use. The process of segregating and reusing fertile topsoils during mining is critical to the later success of reclamation efforts because a suitable growing medium is essential to the reestablishment of viable plant communities. Reclamation of abandoned mine lands where topsoil was not separated from other overburden is generally more difficult and expensive than reclamation at operating mines.

### Reclamation Activities

Reclamation encompasses three activities: backfilling and grading, replacement of topsoil, and revegetation. Backfilling and grading occur after the mineral has been removed. Overburden is replaced in the mined area to reestablish a stable land surface that is consistent with the surrounding area and compatible with the intended postmining land use. Front-end loaders, heavy trucks, bulldozers, and graders are used to fill and grade the contours of highwalls, overburden piles, and depressions to approximate original slopes. The resulting surfaces must be stable—that is, not prone to landslides or erosion—and should blend in with the surrounding natural topography.

Mining operations sometimes unearth natural materials that are toxic or acid-forming. When exposed to the atmosphere, these materials may alter the acidity of surrounding soil or water bodies, destroying the organisms that live there. Such materials must be isolated from surface water and groundwater, soils, and vegetation so that cannot contaminate the environment. This generally means placing them below the root zone of plants during backfilling and grading. During backfilling and grading, heavy equipment repeatedly crosses the work area, causing compaction of the ground surface. Prior to redistribution of topsoil, it may be necessary to rip up this surface to relieve compaction. This helps prevent slippage of topsoil by creating a roughened surface and aids root penetration by vegetation, thus improving surface stability.

After backfilling and grading are completed, a layer of topsoil is spread over the graded overburden to a depth determined by the intended postmining land use and the amount of topsoil available, often 1.2 meters (4 feet) or more. Topsoil stockpiled for more than two or three years begins to lose nutrients, beneficial bacteria, and fungi that aid in plant establishment, so soil tests are used to determine what soil amendments may be needed. Where nutrients are lacking, they are replenished using fertilizers similar to those used on home lawns and gardens.

Revegetation must occur soon after placement of topsoil to control the effects of wind and water erosion. A fast-growing annual grass or cereal grain cover crop, as well as mulch, may be used to stabilize the soil until the first normal planting season. Shrubs and small trees may also be planted. The goal of revegetation is the establishment of a diverse, permanent vegetative cover of a seasonal variety native to the area, or of a variety that supports the intended postmining land use.

In the eastern United States, where water is plentiful, it may take five years to determine whether a mine reclamation project has been successful; in the semiarid western United States, determination may require ten years. Common uses of reclaimed mined land include cropland agriculture, commercial forestry, recreational areas such as parks, public works (such as airfields, roads, housing developments, and industrial sites), and fish and wildlife conservation.

### Reclamation Laws

In the United States, reclamation of land disturbed during the mining of coal has been required since 1977 under national legislation known as the Surface Mining Control and Reclamation Act (SMCRA). Principal responsibility for enforcing this law rests with the U.S. Department of the Interior's Office of Surface Mining Regulation and Enforcement and state regulatory authorities, with programs approved under the statute. A small fund is available for the reclamation of abandoned mine lands, financed by fees on each ton of coal produced by active mining operations.

The requirements of SMCRA apply only to coal mines operating since 1977. Lands disturbed by the mining of gold, silver, nickel, copper, bauxite, limestone, and other minerals and industrial materials are not subject to uniform national standards but may be subject to requirements for reclamation imposed by some states; such state regulations vary greatly, however. The knowledge and technology necessary for successful reclamation of land disturbed by mining are available for almost all ecological systems, except desert and alpine climate conditions that do not favor the rapid plant growth necessary to stabilize reclaimed soil.

*Michael S. Hamilton*

### Further Reading

Hossner, Lloyd R., ed. *Reclamation of Surface-Mined Lands.* Boca Raton, FL: CRC Press, 1988.

Otto, James M. "Global Trends in Mine Reclamation and Closure Regulation." In *Mining, Society, and a Sustainable World,* edited by J. P. Richards. New York: Springer, 2009.

Pipkin, Bernard W., et al. "Mineral Resources and Society." In *Geology and the Environment*. 5th ed. Belmont, Calif.: Thomson Brooks/Cole, 2008.

Schor, Horst J., and Donald H. Gray. *Landforming: An Environmental Approach to Hillside Development, Mine Reclamation, and Watershed Restoration*. Hoboken, N.J.: John Wiley & Sons, 2007.

## Naess, Arne

CATEGORY: Philosophy and ethics
IDENTIFICATION: Norwegian philosopher
BORN: January 27, 1912; Oslo, Norway
DIED: January 12, 2009; Oslo, Norway
SIGNIFICANCE: Naess's ideas, in particular his introduction of the concept of deep ecology, have had a great deal of influence on environmental philosophy and activism.

The wide-ranging philosophy and academic thought of Arne Naess were enriched by Naess's extensive experience with mountaineering. His early interests encompassed the philosophy of science, empirical semantics, and the critique of dogmatism. In the late 1960's Naess began to develop an ecological philosophy (or ecosophy) that he called Ecosophy T, which was heavily influenced by the metaphysics of Baruch Spinoza, the social activism of Mohandas Gandhi, and the spiritual insights of Hinduism and Buddhism.

At the core of Ecosophy T is an intuitive identification with all of life and a spontaneous feeling of unqualified value for the flourishing of all things. This intuition involves a wider sense of self that includes all of existence without dissolving individuality. From this comes an inclination toward maximizing biodiversity, pluralism in philosophy and religion, and decentralization in economics and politics. A central but controversial principle that emerges from Ecosophy T is biospherical egalitarianism—that is, the principle that all things have an equal right to live and bloom. Considered by itself as an absolute, this principle leads to crippling ethical paradoxes, but Naess combined it with a recognition of the inevitability of harming other life and the need to develop a hierarchy of obligations.

In 1972 Naess presented a paper titled "The Shallow and the Deep, Long-Range Ecology Movement" in which he described the concept of deep ecology and its radical orientation to the environment. Deep ecology's holism has been criticized as devaluing individuals, but Naess's philosophy does not present a simple monism that places value only on a single whole. Instead, its nondualistic holism includes an affirmation of the reality and value of individuals and relationships.

Central to Naess's method was an insistent questioning of conventional ideas and values, as well as a probing into the root causes of environmental problems. Such inquiry, Naess believed, leads beyond mere policy debates to critical engagement in worldviews, which are rooted in foundational intuitions. While reason cannot establish these intuitions, it can help people deduce general principles and particular beliefs. Similarly, Naess rejected conventional notions of duty and altruism, which are based on an atomistic view of independent selves. Naess saw morality as grounded in the intuitive experience of identification with others and a spontaneous sense of caring for the well-being of all things. Naess was a phenomenologist; he emphasized a direct experience of reality that goes beyond the subject-object duality. One result of this emphasis on the direct intuition of nature is a pervasive sense of joy.

Naess was not concerned simply with articulating his own ecosophy. True to his antidogmatism, he presented deep ecology as an open-ended philosophical and social movement. Along with George Sessions, Naess created the deep ecology platform, a list of eight principles on which radical environmentalists with different philosophies can agree and that they can use as the basis for activism. Naess championed a diversity of worldviews and beliefs while arguing for the need for solidarity in engaging environmental problems.

*David Landis Barnhill*

FURTHER READING

De Steiguer, J. E. "Arne Naess and the Deep Ecology Movement." In *The Origins of Modern Environmental Thought*. Tucson: University of Arizona Press, 2006.

Katz, Eric, Andrew Light, and David Rothenberg, eds. *Beneath the Surface: Critical Essays in the Philosophy of Deep Ecology*. Cambridge, Mass.: MIT Press, 2000.

# National forests

CATEGORIES: Forests and plants; resources and resource management
DEFINITION: Forestlands owned and managed by a national government
SIGNIFICANCE: The U.S. national forest system was one of the first and most successful resource management and conservation programs initiated by the federal government. Similar programs have been developed in most of the forested nations of the world to manage the forests' economic and ecological functions.

National forests are distinctly different from national parks. National parks are aesthetically pleasing, culturally or historically significant, or ecologically important tracts of land protected in perpetuity for the public benefit. National forests are publicly owned resource management areas operated to maximize the cash value of timber and forest products, benefit the domestic economy, and ensure perpetual productivity of the land.

In the United States, the nineteenth century saw increased efforts to preserve timber resources at the same time that unrestrained national growth encouraged increased timber harvesting. In 1827 U.S. president John Quincy Adams expressed interest in developing a plan for sustainable forestry to ensure the availability of masts for ships. The American Association for the Advancement of Science discussed the need for sustained-yield forestry during the 1860's and in 1873 asked Congress to preserve and manage the nation's forests.

The first congressional actions to preserve forests focused on areas of significant natural beauty, such as Yellowstone National Park and Yosemite Valley. The Division of Forestry was established within the U.S. Department of Agriculture in 1876 to promote harvestable timber development. Protections for harvestable stands of timber were first approved in the Creative Act of 1891, which authorized the president of the United States to withdraw public lands previously open to preemption and homesteading rights and to establish forest reserves.

President Benjamin Harrison established America's first national forest, now known as Shoshone National Forest, in Wyoming in 1891. Confusion arose over the process for including land within the national forest system, however, and Congress suspended the law while it debated the purpose of the reserves, eventually authorizing selective cutting and marketing of timber from the existing reserves. Controversy concerning the vesting of the forest resources in the Department of the Interior or the Department of Agriculture ended in 1905 when the forest reserve system was established. President Theodore Roosevelt vested responsibility for national forest resources in the Bureau of Forestry of the Department of Agriculture under Gifford Pinchot, chief of the Division of Forestry. The reserve system became the national forest system in 1907.

## A Resource with Many Uses

From its inception, the U.S. national forest system was intended to serve multiple purposes. The purposes of the original legislation were to improve water flows and furnish a continuous supply of timber. This evolved into the triple purposes of resource protection (especially fire protection), "wise use" of timber resources, and multiple use of system lands. The forest service lands are protected, harvested, and open to multiple other public uses. The Multiple Use-Sustained Yield Act of 1960 defined "multiple use" as a combination of outdoor recreation, fish and wildlife management, and timber production intended to meet the needs of the American people but not necessarily giving the maximum dollar value return.

The National Forest Management Act of 1976 increased citizen involvement in decisions concerning timber harvesting, resource conservation, and multiple uses. The 1976 act also limited the technique of timber harvesting by clear-cutting, a practice condemned by most environmentalists. In addition, the act limited logging on fragile lands and encouraged actions to maintain the diversity of plants and animals and to conserve plants, animals, soils, and watersheds in the forests. However, the act also emphasized the importance of multiple uses such as mining, oil and gas exploration, grazing, farming, hunting, recreation, and logging. The conflicts among conservation, harvesting, and multiple uses have continued to plague forest policy decision makers. Emerging forest policies in the early twenty-first century include sustained-yield forestry; substitution of alternative harvestable crops for harvestable timber, including crops such as nuts, fruits, gums, extracts, syrups, tars, and oils; and ecosystem maintenance.

The federal government owns and manages approximately 266 million hectares (657 million acres) of public land, about 29 percent of the land area of the United States. Almost 30 percent of the federal land area is part of the national forest system. In 2010, the national forest system comprised roughly 78 million hectares (193 million acres) of land in 155 national forests and 20 national grasslands located in forty-four states, Puerto Rico, and the Virgin Islands. Two-thirds of U.S. national forests are located in the West, the Southeast, and Alaska. Twenty-three eastern states share about 50 forests. Roughly one out of every six hectares in the national forest system is part of the National Wilderness Preservation System (NWPS), and forest system lands represent about 33 percent of the NWPS.

## Other National Forest Programs

Around the world, 143 countries have forest policy statements, and 156 countries have specific forest laws. Almost 75 percent of the world's forests are covered by national forest programs, and about 80 percent are under public ownership.

According to 2010 statistics from the United Nations Food and Agriculture Organization, the world's total forest area is just over 4 billion hectares (9.9 billion acres), or approximately 31 percent of the globe's total land area. Forests in the Russian Federation, Brazil, Canada, the United States, and China account for more than half of the world's forested area. During the 1990's global forests declined at a rate of about 16 million hectares (39.5 million acres) per

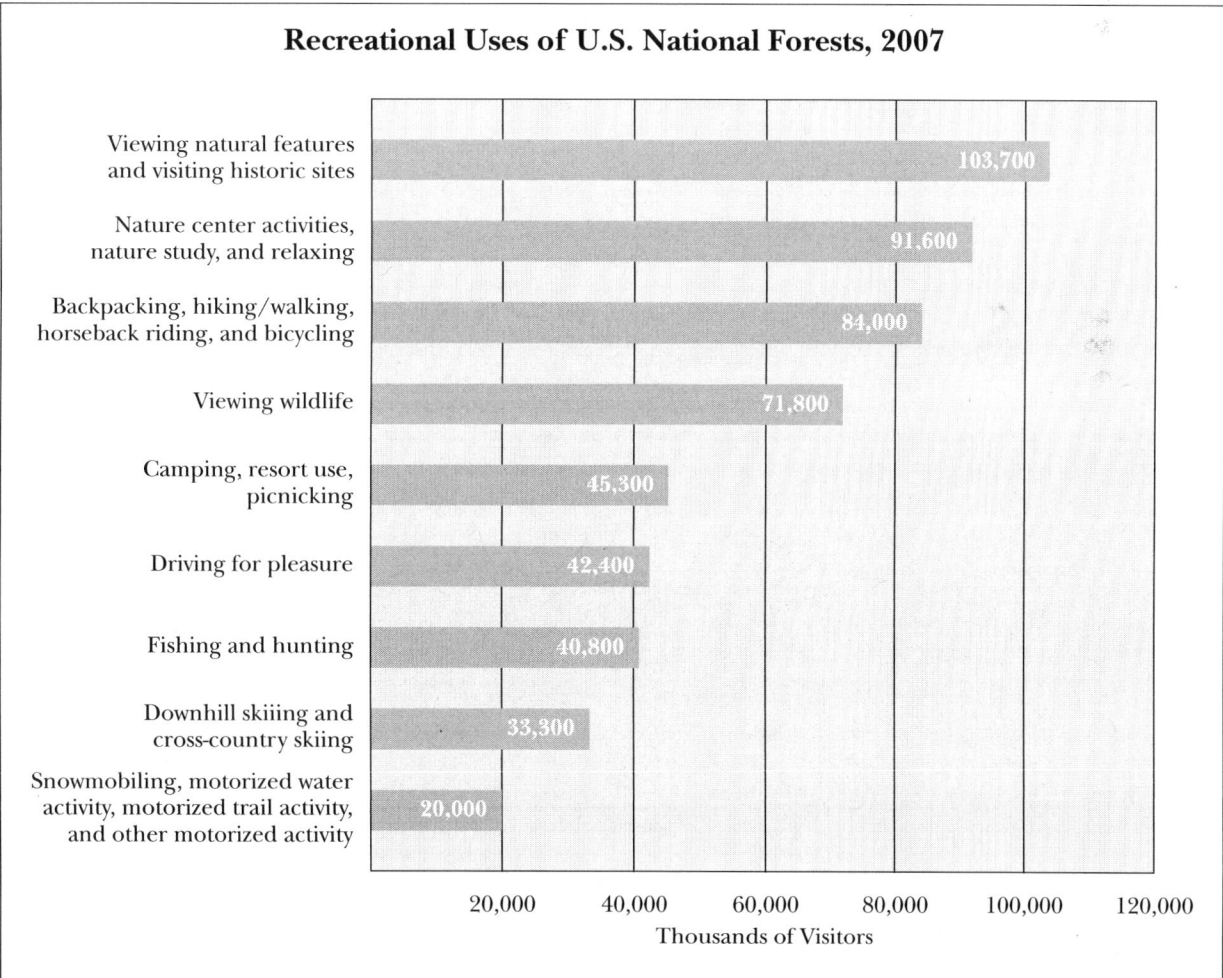

*Source: National Visitor Use Monitoring Results, National Summary Report.* United States Department of Agriculture, Forest Service, 2009.

year. The decline slowed to 13 million hectares (32 million acres) per year between 2000 and 2010. In addition, during this later period natural forest expansion in combination with tree-planting programs added 7 million hectares (17.3 million acres) of new forest annually.

Between 1990 and 2010, the world forest area designated primarily for wood and nonwood forest products decreased, while the area designated for multiple uses increased. By 2010 about 30 percent of the world's forests were designated for productive purposes, 24 percent for multiple uses, and 8 percent for protective functions such as soil and water conservation, avalanche control, sand dune stabilization, and desertification control.

The global area of forested lands in legally protected areas such as national parks and wilderness areas increased between 1990 and 2010 by more than 94 million hectares (232 million acres). In 2010 legally protected areas accounted for about 13 percent of the world's total forest area. Environmental activism, ecotourism, and debt-for-nature swaps encourage the continuing development of national forest systems and other conservation measures in most nations of the world.

*Gordon Neal Diem*
*Updated by Karen N. Kähler*

FURTHER READING

Hayes, Tanya, and Elinor Ostrom. "Conserving the World's Forests: Are Protected Areas the Only Way?" *Indiana Law Review* 38, no. 3 (2005): 595-618.

Hays, Samuel P. *The American People and the National Forests: The First Century of the U.S. Forest Service*. Pittsburgh: University of Pittsburgh Press, 2009.

_____. *Wars in the Woods: The Rise of Ecological Forestry in America*. Pittsburgh: University of Pittsburgh Press, 2007.

Hirt, Paul W. *A Conspiracy of Optimism: Management of the National Forests Since World War Two*. Lincoln: University of Nebraska Press, 1994.

United Nations Food and Agriculture Organization. *Global Forest Resources Assessment, 2010*. Rome: Author, 2010.

# National Park Service, U.S.

CATEGORIES: Organizations and agencies; preservation and wilderness issues
IDENTIFICATION: Agency within the U.S. Department of the Interior responsible for managing national parks, monuments, historic sites, battlefields, recreational areas, and heritage areas
DATE: Established on August 25, 1916
SIGNIFICANCE: The U.S. National Park Service is a world leader in the preservation and protection of natural and cultural resources and has served as a model for the management of national park systems worldwide.

The U.S. Congress established the National Park Service in 1916 to protect natural and historic resources for the enjoyment of future generations, making sites that would otherwise have been lost to development or neglect accessible to the general public. Units within the national park system range from parks covering many thousands of hectares, such as Grand Canyon National Park, to sites consisting of only one building, such as Thaddeus Kosciuszko National Memorial in Philadelphia. Although the designation "national park" has a specific legal meaning, for convenience employees of the National Park Service often refer to all management units within the system as "parks" regardless of size.

A number of federal preserves were designated prior to creation of the National Park Service, such as Hot Springs in Arkansas and Yosemite in California, but management of those sites was split among several different cabinet departments, including the War Department, which detailed troops to patrol Yellowstone and Yosemite. Advocates for wilderness preservation and a uniform national parks policy, such as John Muir, argued that oversight of national preserves should be the responsibility of one agency devoted solely to that task. In 1906, under President Theodore Roosevelt, the U.S. Forest Service had been established within the Department of Agriculture to manage the nation's forest reserves for wise use; promoters of a park service believed a similar agency should be established to manage the nation's parks and monuments.

At the time the Park Service was established, the system consisted primarily of sites noted for their spectacular natural beauty and unique features, such as Yellowstone and Yosemite. However, the 1916 legislation stated the mission of the National Park Service would be "to conserve the scenery and the natural and historic objects and the wild life therein and to provide for the enjoyment of the same in such manner and by such means as will leave them unimpaired for the enjoyment of future generations." This allowed the national park system to include sites important for their historic associations, such as Valley Forge in Pennsylvania, or their archaeological value, such as Mesa Verde in Colorado.

### RESPONSIBILITIES

In the twenty-first century, the U.S. National Park Service, globally recognized as a leader in the preservation of both natural and historic resources, is responsible for managing sites and cultural artifacts in all fifty states, the District of Columbia, Puerto Rico, the U.S. Virgin Islands, Guam, the Northern Marianas, and American Samoa. Sites managed by the Park Service are categorized as national parks, national preserves, national scenic rivers, national recreational areas, national monuments, national historical parks, national historic sites, and national parkways. In addition, the Park Service manages some national trails, administers the National Register of Historic Places and National Historic Landmark programs, and plays an advisory role in the management of National Heritage Areas.

For administrative purposes, the National Park Service is divided into seven regions: Northeast, National Capital, Southeast, Midwest, Intermountain, Pacific West, and Alaska. Each regional office is staffed by experts in cultural and natural resource management, such as wildlife biologists, to provide specialized support to the individual parks within the region. The director of the National Park Service is appointed by the president, but most other employees are members of the civil service.

### DESIGNATION OF SITES

National monuments can be created by presidential executive order, but congressional action is required to create sites in all the other categories under the Park Service. The executive branch's power to protect unique natural and cultural resources dates back to President Theodore Roosevelt and the 1906 Antiquities Act, legislation that was passed to allow the president to provide immediate protection to threatened resources. The fact that a site has been designated as a unit of the national park system, whether through congressional action or by executive order, does not guarantee permanent protection of that site. Congress can remove any unit in the system by passing legislation to do so. This rarely happens; in general, once a site becomes part of the national park system, it is more likely that the site will be enhanced and expanded rather than the reverse.

The history of the Park Service and the expansion of the national park system are reflections of U.S. history overall and of changes in thinking among Americans regarding what is worthy of preservation. Since establishment of the National Park Service in 1916, numerous sites throughout the United States have been proposed as additions to the system. When a site is suggested, Congress first authorizes a study to determine whether the proposed unit meets the criteria for its intended designation—for example, whether a historic site's significance is truly of national importance. Creation of a national park can be an intensely political process, with vigorous lobbying for and against the proposal. In addition, the vision of what a national park should be has changed over time and continues to evolve, so that sites that may have been rejected in the past may eventually come to be seen as appropriate.

In addition, partnerships between the Park Service and nonfederal agencies and organizations are becoming more common, as is seen in parks such as the Lewis and Clark National and State Historical Parks in Washington and Oregon and Keweenaw National Historical Park in Michigan. Both parks include some

### National Park System Statistics, 2000-2007

|  | 2000 | 2005 | 2007 |
| --- | --- | --- | --- |
| Expenditures | $1.833 billion | $2.451 billion | $2.412 billion |
| Revenue from operations | $234 million | $286 million | $346 million |
| Recreational visitors | 285,900,000 | 273,500,000 | 275,600,000 |
| Overnight stays | 15,400,000 | 13,500,000 | 13,800,000 |
| Park system lands (acres) | 78,153,000 | 79,048,000 | 78,845,000 |

*Source:* U.S. Census Bureau, *Statistical Abstract of the United States, 2009,* 2009.
*Note:* Includes visitor data for national parks, monuments, recreation areas, seashores, and miscellaneous other areas.

sites owned and managed directly by the National Park Service and some sites owned and managed by local government agencies or nonprofit organizations.

*Nancy Farm Männikkö*

FURTHER READING

Albright, Horace M., and Marian Albright Schenk. *Creating the National Park Service: The Missing Years.* Norman: University of Oklahoma Press, 1999.

Duncan, Dayton, and Ken Burns. *The National Parks: America's Best Idea—An Illustrated History.* New York: Alfred A. Knopf, 2009.

Farabee, Charles R. *National Park Ranger: An American Icon.* Lanham, Md.: Roberts Rinehart, 2003.

Heacox, Kim. *An American Idea: The Making of the National Parks.* Washington, D.C.: National Geographic Society, 2001.

# National parks

CATEGORY: Preservation and wilderness issues
DEFINITION: Areas of scenic, historic, or other value that are set aside by federal governments for the preservation of animals and wildlife and for human recreation
SIGNIFICANCE: National park systems throughout the world provide protection for wildlife and plant life, preserve spaces for outdoor recreation, and educate people about the importance of protecting ecosystems.

National parks are places where preservation for future generations must be balanced with present-day enjoyment. In the United States this balance has proved difficult to achieve, but nevertheless more than 390 park units—including national parks, historic sites, historical parks, memorials, memorial parks, battlefields, battlefield parks, battlefield sites, lakeshores, seashores, monuments, parkways, scenic trails, scenic rivers, scenic riverways, rivers, capital parks, and recreation areas—have been established across the United States since the U.S. Congress made Yellowstone the world's first national park in 1872 (the politicians were willing to protect and preserve the geologic wonders of Yellowstone primarily because they were convinced that the lands were economically useless). Each of these park units was established to preserve and protect geologic wonders, spectacular scenery, wildlife, or a particular aspect of American history or culture.

Other countries have also established national parks, frequently using the United States as a model. Starting in the final decades of the twentieth century, the United Nations worked with countries to protect these areas. In 1972, on the occasion of the one hundredth anniversary of the founding of Yellowstone National Park, the United Nations Educational, Scientific, and Cultural Organization (UNESCO) formed the World Heritage Committee. By 2009, 186 countries had ratified the World Heritage Convention. The committee, at its first meeting in 1977, formulated the World Heritage List, which names cultural and natural sites considered to be of "outstanding universal value"; approximately thirty new sites have been added to the list each year. By the end of the first decade of the twenty-first century, the World Heritage List named almost nine hundred cultural and natural sites. The list contains more cultural than natural sites, and not all the sites are parks, but many parks have received monetary support and advice from the United Nations to help ameliorate environmental problems, which abound in the United States and the rest of the world; the causes of these problems include the difficulties posed by the poaching of wildlife, exploitation of mineral deposits, and finding a balance between use and preservation.

NATIONAL COORDINATING OFFICES

Initially, national park operations in the United States were complicated because no single central federal office existed to coordinate activities. After much controversy, Congress passed the National Parks Act, or the Organic Act, of 1916. The act established a central authority called the National Park Service and stated its responsibilities, which include conserving and providing for the enjoyment of the scenery, natural and historic objects, and wildlife in the parks while leaving them unimpaired for the enjoyment of future generations.

This was not the first national office of national parks. In 1911 the Canadian parliament had passed the Dominion Forest Reserves and Parks Act, which provided for the administration of forest reserves and dominion parks, and allowed dominion parks to be established from forest reserves. Thus the Dominion Parks Service, created as a new branch in the Canadian Department of the Interior, became the first distinct bureau of national parks in the world. John Ber-

*Mammoth Hot Springs in Yellowstone National Park, the world's first national park.* (Courtesy, PDPhoto.org)

nard Harkin served as commissioner from the service's inception in 1911 to 1936. In addition to working to separate the administration of the parks from that of the forests, Harkin emphasized resource preservation.

The numbers of national parks, as well as the numbers of visitors to the parks, continued to grow in both the United States and Canada. After World War II, as the automobile became ubiquitous, visits to the national parks increased rapidly. The facilities in place at the parks were old, however, and staffing was minimal. In 1956 the director of the U.S. National Park Service, Conrad L. Wirth, decided that instead of asking Congress for annual appropriations, he would package the national parks' needs into a program called Mission 66. This one-billion-dollar, ten-year restoration program was designed to end in 1966, the fiftieth anniversary of the National Park Service. Canada was also influenced by Mission 66, and the Parks Policy of 1964, under the leadership of John I. Nicol, director of the Canadian National and Historic Parks Branch, emphasized the importance of protecting natural resources in the parks.

## Problems of Overuse

The numbers of visitors to national parks in both countries increased steadily from the 1960's onward. With larger crowds came overuse, which created its own serious problems, including congestion, pollution, and—in some cases—destruction. Some popular parks banned automobiles from their roads, replacing them with shuttle buses. The park services of both Canada and the United States are continually faced with the problem of encouraging use while protecting valuable national resources and leaving them unimpaired for future generations.

In the United States, most of the service businesses in the national parks, such as restaurants, hotels, and souvenir shops, have traditionally been operated by private concessionaires. The operators usually sign long-term contracts under which small percentages of their profits are returned to the federal government. In the late twentieth century, as public funding for the parks declined and the numbers of visitors continued to increase, the National Park Service intensified its ties with private agencies to provide public services at national parks and other sites administered by the service. Some proposed projects became controversial and were never implemented, such as a giant theater that was to be built at Gettysburg Battlefield.

The National Park Service recognizes the importance of cooperation with the communities immediately adjacent to the parks, and partnerships and resulting plans have been developed around several national parks. Several of Canada's national parks, such as Banff and Jasper, actually contain towns, and the animals that live in the parks roam freely among automobiles, homes, and stores.

## Management of Plants and Wildlife

Both the Canadian and the U.S. national park services must balance use with preservation of wildlife and plants. In general, the aim is to allow native species to flourish within the parks while protecting the parks from exotic, or nonnative, species. However, conflicts frequently develop regarding the remedial approaches that should be taken when exotic species are found in the parks. Some parks, such as the Shenandoah and Great Smoky Mountains national parks, have been invaded by exotic insects that have destroyed whole forested areas and the component parts of the ecosystem.

One of the most controversial decisions in recent U.S. National Park Service history was the decision to reintroduce the gray wolf into the Yellowstone ecosystem in 1995. Wolves in the area had been exterminated by rangers in 1915 because they were seen as a menace to other native animals such as elk, deer, and mountain sheep. After many years of debate, the National Park Service was given permission and funds (approximately $6.7 million) to reintroduce wolves to the Yellowstone ecosystem, which includes the park and neighboring parts of Montana and Idaho. Many park visitors, upon being surveyed, had indicated that they wanted the wolves brought back. Local livestock owners, however, were concerned about the safety of their animals. To balance the competing needs—to serve the desires of park visitors, to restore the natural ecology of the park, to address the economic concerns of the livestock owners, and to support efforts to improve the wolf population—the Park Service developed plans to protect neighboring livestock before it undertook the careful reintroduction of wolves. In January, 1995, the first set of gray wolves from Hinton, Alberta, Canada, were brought to Yellowstone. The introduction was deemed successful, and the ecology of the park returned to a more natural state as the wolf population increased and the wolves became significant predators of elk, moose, and deer.

One problem area related to wildlife that affects almost all national parks involves the dangers that arise from human-animal interaction. It is illegal for visitors to feed any wildlife in U.S. and Canadian national parks, but some visitors break the law. Human food is not part of the animals' natural diet, and some foods can cause digestive problems in animals and even endanger their lives; in addition, animals that come to associate humans with a ready food source can be dangerous to humans. When wild animals enter campgrounds and motor vehicles in search of something to eat, they occasionally harm humans. In extreme cases, to prevent animals from hurting people, park rangers may have to transport, or even kill, animals that have become too aggressive in searching for humans' food.

## Economic Activities and Environmental Problems

The economic activities taking place outside park boundaries can cause environmental problems inside the parks. In some cases, neighboring mining or logging operations have contaminated waterways or hastened soil erosion. Sometimes compromises are reached that can reduce the impacts of such problems. For example, a planned gold mining operation outside Yellowstone National Park was moved when the U.S. government agreed to exchange other federal lands for the original intended site.

In other cases, distant power plants, factories, or even urban areas as a whole contribute to air pollution within park boundaries. In many parks in the United States, the combination of human-caused pollution and natural geology and meteorology has led to diminished views, more rapid weathering of natu-

> ### "Fixing" the National Parks
>
> *National Park Service director Conrad Wirth presented his plan to "fix" the park system. Wirth's proposal, "Mission 66: Special Presentation to President Eisenhower and the Cabinet" (January 27, 1956), reads, in part:*
>
> As you know, these are the areas where the Nation preserves its irreplaceable treasures in lands, scenery—and its historic sites—to be used for the benefit and enjoyment of the people, and passed on unimpaired to future generations. The areas of the National Park System are among the most important vacation lands of the American people. Today, people flock to the parks in such numbers that it is increasingly difficult for them to get the benefits which parks ought to provide, or for us to preserve these benefits for Americans of tomorrow.
>
> This is why the Department of the Interior and the National Park Service have surveyed the parks and their problems, and propose to embark upon Mission 66—a program designed to place the national parks in condition to serve America and Americans, today and in the future.
>
> The problem of today is simply that the parks are being loved to death. They are neither equipped nor staffed to protect their irreplaceable resources, nor to take care of their increasing millions of visitors.
>
> Where else do so many Americans under the most pleasant circumstances come face to face with their Government? Where else but on historic ground can they better renew the idealism that prompted the patriots to their deeds of valor? Where else but in the great out-of-doors as God made it can we better recapture the spirit and something of the qualities of the pioneers? Pride in their Government, love of the land, and faith in the American Tradition—these are the real products of our national parks.

ral wonders, and harm to wildlife. In South Africa, the industrial district of Saldanha Bay has contributed to air and water pollution in the West Coast National Park.

National parks in parts of Africa and Central America face serious problems of wildlife poaching because of inadequate park administration and law enforcement. In areas such as Benin, disease-carrying livestock sometimes invade parks, resulting in wildlife deaths. Parts of national parks in Côte d'Ivoire and Senegal have been cultivated by farmers, but the governments of both countries have developed successful resettlement programs, thus lessening the impact on the parks.

Individual countries often look for help from the World Heritage Committee in establishing and maintaining national parks. The committee keeps a list of sites known as World Heritage in Danger, with the intent of bringing to worldwide attention the "conditions which threaten the very characteristics for which a property was inscribed on the World Heritage List." In 2010, thirty-one sites were on the World Heritage in Danger list. Sometimes, successful intervention results in a site being removed from the list, as occurred in 1998 with Plitvice Lakes National Park in Croatia. The park had been overdeveloped and overused, but after it was added to the World Heritage in Danger list in 1992, its underground water supply was protected and a new road was built to decrease truck traffic through the park.

*Margaret F. Boorstein*

FURTHER READING

Allin, Craig W., ed. *International Handbook of National Parks and Nature Reserves.* New York: Greenwood Press, 1990.

Grusin, Richard. *Culture, Technology, and the Creation of America's National Parks.* New York: Cambridge University Press, 2004.

Heacox, Kim. *An American Idea: The Making of the National Parks.* Washington, D.C.: National Geographic Society, 2001.

Ridenour, James. *The National Parks Compromised: Pork Barrel Politics and America's Treasures.* Merrillville, Ind.: ICS Books, 1994.

Runte, Alfred. *National Parks: The American Experience.* 4th ed. Lanham, Md.: Taylor Trade, 2010.

Sellars, Richard. *Preserving Nature in the National Parks: A History.* New ed. New Haven, Conn.: Yale University Press, 2009.

## National Trails System Act

CATEGORIES: Treaties, laws, and court cases; preservation and wilderness issues
THE LAW: U.S. federal law authorizing a national network of scenic, historic, and recreational trails protected from intrusive development
DATE: Enacted on October 2, 1968
SIGNIFICANCE: Hundreds of thousands of people use

the system of protected trails created by the National Trails System Act every year to gain access to scenic, historic, and culturally important sites across the United States.

When the Appalachian Trail, a hiking trail running from Georgia to Maine, was developed during the 1920's and 1930's, it was intended to provide a wilderness experience for hikers from the cities. By the 1960's, however, that wilderness was threatened by commercial and residential development, with its accompanying roads and utilities. Supporters of the Appalachian Trail lobbied Congress to protect the trail, and in 1965 President Lyndon B. Johnson issued his "Special Message to the Congress on Conservation and Restoration of Natural Beauty," in which he called for a nationwide system of trails that would be largely maintained by volunteers. The next year, Secretary of the Interior Stewart Udall directed the Bureau of Outdoor Recreation to study the issue, and the result was an influential report titled *Trails for America*, which outlined plans for a system of trails throughout the country.

The National Trails System Act, as enacted in 1968, defined and created three different types of trails as described in *Trails for America*: national scenic trails, national recreation trails, and connecting and side trails. National scenic trails are trails and protected corridors that are more than 161 kilometers (100 miles) long, linking points of extraordinary scenery. Motorized vehicles are forbidden on these trails, to protect the wilderness experience, and only rudimentary development, such as campsites or shelters, is permitted. Only Congress can create national scenic trails, and the responsibility for maintaining them falls to the Department of the Interior. Congress has created eight such trails; these include the Appalachian National Scenic Trail and the Pacific Crest National Scenic Trail, both established in 1968 with the signing of the act. No new national scenic trails have been designated since 1983.

In 1978, Congress amended the National Trails System Act to add national historic trails, which mark historically important routes of travel. The Iditarod Trail, the Lewis and Clark Trail, the Mormon Pioneer Trail, and the Oregon Trail were designated as the first in this group; by 2009 the list had grown to include nineteen trails.

National recreation trails are generally smaller, more locally controlled, and more open to a variety of uses, including the riding of motorized vehicles. Designation of these trails, which number more than eight hundred, is not subject to congressional approval, and the trails do not receive federal funding. Side and connecting trails are trails that connect to other trails in the National Trails System; although such trails are provided for in the original act, only two have been established.

The National Trails System has proved to be a great success, attracting thousands of visitors each year to places of scenic and historic importance. Many of the trails are still maintained by volunteers, as President Johnson originally envisioned. Each year, on the first Saturday in June, thousands of organizations and businesses observe National Trails Day with hikes, horseback rides, mountain bike rides, trail-grooming events, and educational workshops.

*Cynthia A. Bily*

---

### National Trails System Act

*The National Trails System Act of 1968 was landmark legislation that provided expanded outdoor recreation opportunities at the same time that it ensured protection of the nation's historic trails system. The act, in part, states:*

In order to provide for the ever-increasing outdoor recreation needs of an expanding population and in order to promote the preservation of, public access to, travel within, and enjoyment and appreciation of the open-air, outdoor areas and historic resources of the Nation, trails should be established (i) primarily, near the urban areas of the Nation, and (ii) secondarily, within scenic areas and along historic travel routes of the Nation which are often more remotely located.

The purpose of this Act is to provide the means for attaining these objectives by instituting a national system of recreation, scenic and historic trails; by designating the Appalachian Trail and the Pacific Crest Trail as the initial components of that system; and by prescribing the methods by which, and standards according to which, additional components may be added to the system.

The Congress recognizes the valuable contributions that volunteers and private, nonprofit trail groups have made to the development and maintenance of the Nation's trails. In recognition of these contributions, it is further the purpose of this Act to encourage and assist volunteer citizen involvement in the planning, development, maintenance, and management, where appropriate, of trails.

FURTHER READING

Baldwin, Pamela. *Federal Land Management Agencies.* Hauppauge, N.Y.: Nova, 2005.

Dilsaver, Lary M., ed. *America's National Park System: The Critical Documents.* Lanham, Md.: Rowman & Littlefield, 1997.

Johnson, Sandra L. "The National Trails System: An Overview." In *National Forests: Current Issues and Perspectives,* edited by Ross W. Gorte. Hauppauge, N.Y.: Nova, 2003.

# Nature Conservancy

CATEGORIES: Organizations and agencies; preservation and wilderness issues

IDENTIFICATION: American nonprofit organization that works to preserve threatened ecosystems and the plants and animals that inhabit them

DATE: Founded in 1951

SIGNIFICANCE: By purchasing land and by helping other landowners to develop conservation plans, the Nature Conservancy has been successful in protecting unspoiled areas from development and pollution.

The Nature Conservancy, based in Arlington, Virginia, traces its origins to a committee of the Ecological Society of America, which was founded in 1915. In 1946, the committee became the Ecologists' Union, which changed its name to the Nature Conservancy in 1950 and was incorporated as a nonprofit organization in 1951. The group, which included research scientists and conservation advocates, began acquiring land through donation and purchase and collaborating with other agencies to manage public lands to sustain biodiversity.

During the 1960's the Nature Conservancy helped create a new tool for encouraging conservation, the conservation easement, which enables private landowners to receive tax benefits for conserving their land and restricting development while retaining ownership. By 2010 the Nature Conservancy had

*The Nature Conservancy and the U.S. National Park Service have undertaken a joint effort at Channel Islands National Park in California to protect the tiny island fox, a species found nowhere else in the world.* (NPS)

more than one million members and branches in all fifty U.S. states, as well as in more than thirty other countries. It owned more than 405,000 hectares (1 million acres) of land in the United States and approximately the same amount total in other countries, and was involved with other landowners in hundreds of projects addressing the conservation of water, forests, and marine habitats, as well as issues related to climate change.

The Nature Conservancy partners with corporations, government agencies, and other environmental groups, as well as with private landowners. Most of its funding comes from private donations, although some of its experimental projects are supported by government grants. The organization has helped to create and preserve several important areas throughout the United States, including the Golden Gate National Recreation Area in California, the Tallgrass Prairie Preserve in Oklahoma, the Great Sand Dunes National Park and Preserve in Colorado, and Glacial Ridge National Wildlife Refuge in Minnesota. It also conducts programs in other countries and has helped organize debt-for-nature swaps, under which developing nations have some of their international debt canceled in exchange for conserving tracts of ecologically valuable land. In addition to working internationally on an increasingly large scale, the Nature Conservancy encourages its many members to participate in efforts such as the Plant a Billion Trees Campaign, which plants one tree in Brazil for every dollar donated, and its Adopt an Acre program, which allows donors to designate particular habitats to be protected through their donations.

The Nature Conservancy is the wealthiest environmental group in the world, with more than one billion dollars in assets, and its large size and tremendous influence have created controversy. Critics have charged that some of the organization's acquisitions and logging, drilling, and mining projects, while intended to be models of sustainable development, have ignored the rights of indigenous peoples or threatened endangered species. The group's advocacy of integrated fire management has also been controversial. In addition, the Nature Conservancy has been criticized for acquiring private lands through charitable gifts and then reselling them against the wishes of donors, often at large profit to the U.S. federal government.

*Cynthia A. Bily*

FURTHER READING

Birchard, Bill. *Nature's Keepers: The Remarkable Story of How the Nature Conservancy Became the Largest Environmental Organization in the World.* New York: John Wiley & Sons, 2005.

Brewer, Richard. *Conservancy: The Land Trust Movement in America.* Lebanon, N.H.: University Press of New England, 2004.

## Nature preservation policy

CATEGORY: Preservation and wilderness issues
DEFINITION: Decisions made and regulations put in place by any level of government to undertake the protection of natural resources
SIGNIFICANCE: Governments generally promote their policies concerning nature preservation through the passage of legislation. In the United States, changes in nature preservation policy over time have reflected the changes that have taken place in the attitudes of the public toward the need for government protection for the natural environment.

The Industrial Revolution, which diminished traditional agriculture while encouraging urbanization and technology, began straining the relationship between humanity and natural resources during the early nineteenth century. As the technological advances being made in Europe rapidly spread west, environmental damage and natural resource depletion escalated in the United States, inspiring the American conservation movement. The conservation and environmental movements have continued to exert tremendous influence on policy making.

American artist George Catlin first proposed setting aside land for wildlife and Native Americans during the nineteenth century, and in 1864 geographer George Perkins Marsh published *Man and Nature: Or, Physical Geography as Modified by Human Action*, the first influential book to address the human impact on nature. The Homestead Act of 1862 greatly encouraged expansion in the western United States by giving more than 405 million hectares (1 billion acres) of land to settlers, a policy that often resulted in barren landscapes. Destructive logging methods were employed, land was rapidly cleared for agriculture, and large-scale fires raged. Some western grasslands experienced such excessive grazing that many regions had not recovered their full productivity by the end of the twentieth century.

As farmers' journals described "wearing out" several homesteads during westward journeys, naturalist Henry David Thoreau and essayist Ralph Waldo Emerson countered with writings that fueled increasing support for nature conservation. The public expressed considerable outrage regarding the near elimination of several wildlife species that previously had existed in massive numbers, such as bison, deer, elk, and beaver. This led to legislation that created the world's first national park in 1872 at Yellowstone, Wyoming, followed by an 1873 petition to Congress by the American Association for the Advancement of Science to curtail the inefficient use of natural resources such as water, soil, forests, and minerals.

The 1891 Forest Reserve Act began the establishment of natural forests, and the 1900 Lacey Act initiated wildlife protection by regulating commercial hunting. The 1894 Buffalo Protection Act provided recognition that a previously abundant natural resource could rapidly become an endangered species. As naturalist John Muir championed numerous wilderness preservation projects and became the first president of the Sierra Club in 1892, federal legislation in the United States began classifying natural resources as renewable or nonrenewable. Renewable resources can be regenerated and even improve under proper management but can be depleted or completely eliminated if misused. Nonrenewable resources are present only in fixed amounts and will not regenerate regardless of human efforts. Examples of renewable resources include plants, animals, soils, and inland waters; nonrenewable resources include minerals and fossil fuels. The founding of private conservation organizations such as the American Forestry Association in 1875, the American Ornithologists' Union in 1883, the Boone and Crockett Club in 1887, and the New York Zoological Society in 1895 increased public influence on conservation legislation at all levels of government.

### THEODORE ROOSEVELT AND FRANKLIN D. ROOSEVELT

President Theodore Roosevelt initiated habitat protection for wildlife in 1903 when he set aside Pelican Island in Florida's Indian River as a federal bird sanctuary. Through such initiatives as the 1908 White House Governors' Conference on Conservation, Roosevelt's politics and personality helped establish more than fifty wildlife refuges, five national parks, and eighteen national monuments, and increased the area of national forests by more than 69.7 million hectares (150 million acres). Roosevelt's policies required that certain public lands be held in trust for the "good of the country" and separated many public domain regions from commercial interests.

During and immediately following Theodore Roosevelt's presidency, Forest Service chief Gifford Pinchot and Interior Secretary James Garfield implemented more unified policies governing natural resource planning that relied on scientific principles, leading to development of the discipline of conservation biology. Many nonrenewable resources were then protected from exploitation by private industry by the 1920 Mineral Leasing Act.

Political debates and administration changes during the Great Depression of the 1930's shelved more environmental legislation until President Franklin D. Roosevelt signed the Taylor Grazing Act in 1934, whereby all public domain lands would be managed as part of the public trust. The dry Dust Bowl years of the Great Plains states during the early 1930's severely depleted migratory bird populations, motivating a renewed surge of public conservation activity and passage of the 1934 Duck Stamp Act, which tacked a conservation fee for the acquisition of wetlands onto waterfowl hunting licenses.

In 1933 the Soil Erosion Service (later called the Soil Conservation Service), the Civilian Conservation Corps, and the Tennessee Valley Authority were established to provide water and soil conservation assistance to landowners as farmland in the Midwest continued to deteriorate under improper agricultural practices. Franklin Roosevelt's Civilian Conservation Corps provided unemployed Americans with more than two million jobs planting trees and building irrigation systems and dams. More federal involvement was initiated after dust clouds from the dry soil of midwestern farmland blew east all the way to Washington, D.C. In 1940 the U.S. Congress enacted the Bald Eagle Protection Act to protect the national bird.

### POST-WORLD WAR II CONSERVATION

Advances in technology and economic development in the turbulent era following World War II, combined with the postwar baby boom, put additional stressors on environmental resources. President Harry Truman began a national program for water-pollution control, with later legislation requiring states to set and enforce standards for natural rivers. In attempts to reduce insect-borne disease and increase food production, dichloro-diphenyl-trichloroethane (DDT) and other synthetic pesticides were developed and had considerable initial success, causing the near-complete disappearance of malaria and the production of bumper crops.

The 1947 Forest Pest Control Act provided for the detection and chemical destruction of insects that carried diseases harmful to humans, but the numerous new and powerful experimental substances being invented caused other severe environmental problems, which in many cases caused more damage than those that the pesticides were created to prevent. Grassroots public outcry stimulated several federal restrictions on chemicals such as DDT, with many citizens alerted to these dangers by former U.S. Fish and Wildlife Service biologist Rachel Carson's 1962 book *Silent Spring*. Carson is credited with warning mainstream America about the health and environmental hazards posed by pesticides and other toxic chemicals. Her work stimulated further writings that described human threats to the environment, including *The Population Bomb* (1968), by Paul R. Ehrlich, and *The Limits to Growth* (1972), by Donella H. Meadows, Dennis L. Meadows, Jørgen Randers, and William W. Behrens III.

All forms of environmental pollution greatly increased during the 1950's and 1960's. Television beamed graphic examples of environmental problems into public view, notably the mercury poisoning at Minamata Bay, Japan; killer smog episodes in London, England, and Los Angeles, California; and the 1967 *Torrey Canyon* oil spill in the English Channel. As the prices of land and water rights skyrocketed, the Land and Water Conservation Fund, which was set up by federal legislation during the 1960's to increase outdoor recreation space in the United States, generated revenues from offshore drilling leases. Several catastrophic environmental events occurred in 1969, including toxic waste fires on the Cuyahoga River in Cleveland, Ohio, and a coastal oil spill near Santa Barbara, California. Public pressure regarding these and other concerns led to passage of the National Environmental Policy Act of 1969 (NEPA), which became law on January 1, 1970. During the development of this precedent-setting act, Congress discovered that more than eighty governmental units had activities directly affecting the environment, but no government policies were in place to coordinate and review such activities.

Private individuals and organizations such as the Sierra Club, the Nature Conservancy, the Wilderness Society, the National Wildlife Federation, and the National Audubon Society began lobbying for more laws to establish nature preservation areas for both renewable and nonrenewable natural resources. Two highly visible social programs that influenced public opinion were conducted by the Nature Conservancy: Oklahoma's Tallgrass Prairie National Preserve "Adopt a Bison" program and Montana's Pine Butte Swamp Preserve, where dinosaur fossils were discovered in 1978. The Endangered Species Preservation Act of 1966 and the Endangered Species Conservation Act of 1969 did not directly protect any species, but they led to later legislation that did.

## The 1970's and 1980's

Following unanimous passage of NEPA by Congress over President Richard Nixon's objection, the 1970's saw the passage and often complicated enforcement of several laws regulating nature preservation. The Environmental Protection Agency (EPA) was established in 1970, followed later that year by passage of significant amendments to the 1963 Clean Air Act. Important pollution-control measures were then implemented by the 1972 Water Pollution Control Act, the 1973 Endangered Species Act, the 1976 Toxic Substances Control Act, the 1976 National Forest Management Act, and the 1977 Clean Air Act amendments. The Endangered Species Act is considered the most effective and wide-reaching act ever passed by Congress to protect natural ecosystems.

Key legislation supporting nature preservation that was passed during the 1980's included the 1980 National Acid Precipitation Act, the 1980 Alaska National Interest Lands Conservation Act (which enabled the size of the refuge and park systems to double), and the 1987 amendments to the 1972 Clean Water Act. Surveys conducted during the late 1980's revealed that more than 60 percent of wildlife areas in the United States were permitting activities that were harmful to wildlife, with the most destructive practices, such as military activities and drilling, not falling under legal jurisdiction of the U.S. Fish and Wildlife Service. This era also saw increased public interest in nature preservation following events such as the 1986 Chernobyl nuclear power plant catastrophe and the 1989 *Exxon Valdez* oil spill, as well as controversies concerning acid rain, tropical deforestation, the harvesting of old-growth timber, and the discovery of a trend toward global warming.

Private corporations that had previously sacrificed important wildlife habitat began to realize that the environment was an important issue to American consumers. In response to pressure from consumers, employees, and stockholders, many businesses implemented stewardship programs designed to protect natural resources and allow public enjoyment of their underdeveloped lands.

## The 1990's and Beyond

Many nature preservation goals first proposed during the Industrial Revolution finally began to be realized with the systematic creation and maintenance of healthy forests; the prevention of timber depletion and siltation of streams; the provision of food, cover, and protection for wildlife; and the establishment of places where human beings could escape growing urbanization by taking part in outdoor recreation. Water reservoirs could now control flooding, provide clean water for humans and livestock, keep the soil fertile for agriculture, provide irrigation, and generate power. Water treatment plants were now effective in keeping rivers clean by processing wastes from urban sewage, while fish hatcheries provided supplemental stocks to natural and human-made reservoirs, streams, and lakes.

Scenic easements along riverbanks aided antipollution efforts and reduced erosion, while green spaces required by city zoning regulations held soil and became available for community use while maintaining the environment's natural beauty. Mass interurban transportation systems moved people efficiently, and footpaths and bicycle trails offered outdoor recreation and cultural opportunities. Continual management of resources was more successful in keeping delicate ecosystems in balance, with ongoing environmental efforts including the seeding of wildlife foods, controlled burning to destroy unwanted vegetation, and the closing of wildlife habitats during mating and birthing seasons.

Conservationists-turned-environmentalists greatly influenced nature preservation policies during the 1990's as President George H. W. Bush passed legislation in 1990 that amended the 1970 Clean Air Act to focus more on reducing acid rain and emissions from fossil fuels and nitrogen oxide. However, an activist citizens' commission formed in 1992 by the Defenders of Wildlife found that the United States was "falling far short" of meeting the urgent needs of nature preservation. The public response to this information helped lead to passage of the 1997 National Wildlife Refuge System Improvement Act, which shifted the priorities of nature preservation systems toward the formation of multiple-use environments. This key legislation redefined the mission statement regarding conservation of habitats for fish, wildlife, and plants; designated priority public uses such as hunting, fishing, wildlife observation and photography, and environmental education and interpretation; and required that "environmental health" be maintained on public lands. The principle of multiple use, however, continues to allow mining, drilling, grazing, logging, and motorized recreation, as well as military training such as bombing, tank, and troop exercises, on lands designated for nature preservation.

## International Efforts

Cooperative international nature preservation efforts began with the 1918 Migratory Bird Treaty signed by the United States and Great Britain (for Canada), and later Mexico. The International Union for Conservation of Nature and Natural Resources, founded in 1948, represented the interests of 116 countries toward protecting endangered and threatened "living resources." The United Nations Conference on the Human Environment, hosted by Sweden in 1972, was instrumental in establishing nature preservation as an international concern. Utilizing concepts from this conference, the U.S. Congress passed the 1983 International Environmental Protection Act, which included landmark legislation incorporating wildlife and plant conservation and biological diversity as objectives when the United States provides assistance to developing countries.

The 1973 Convention on International Trade in Endangered Species (CITES) involved the cooperation of more than one hundred nations to regulate the import and export of natural resources. Earth Day, first held on April 22, 1970, as a campus-based event encompassing an estimated twenty million people across the United States, began to be combined with other concerns, and annual observances of Earth Day have continued to demonstrate massive support for conservation issues. The international Greenpeace Foundation was formed in 1971 with the aim of applying pressure to governments and organizations to stop such practices as the testing of nuclear weapons and the dumping of radioactive and toxic wastes.

Sentiments favoring the preservation of nature began taking political form in Europe during the early 1980's, notably with the formation of the Green Party in Germany. International collaboration on environmental preservation issues that influenced later legislation included the 1987 Montreal Protocol to protect the ozone layer, the 1992 United Nations Conference on Environment and Development (also called the Earth Summit) in Brazil, and the 1994 United Nations Population Conference in Egypt. The 1992 Earth Summit in Rio de Janeiro, Brazil, which was the largest international meeting ever held (representatives of 178 nations attended), emphasized an approach to nature conservation that focused on sustainable growth and utilitarian solutions. A paper that resulted from the Earth Summit titled "World Scientists' Warning to Humanity" warned that if current consumption rates continued, the earth's resources would be reduced to the point where the world would be "unable to sustain life in the manner we now know."

This was followed by another U.N. summit in 1997 in New York City, attended by representatives of more than fifty nations, during which the progress made in the intervening years was reviewed. Many analysts noted that although progress had been made, the local and global issues surrounding nature preservation and environmental problems were continuing to grow in complexity, and many kinds of environmental damage may be irreversible. Although the attendees did agree to take further action on issues of nature preservation, they made few concrete commitments. The five-year review process continued with the 2002 World Summit on Sustainable Development, which was held in Johannesburg, South Africa; the United States did not participate in this conference, which also produced few concrete commitments.

*Daniel G. Graetzer*

FURTHER READING

Chiras, Daniel D., and John P. Reganold. *Natural Resource Conservation: Management for a Sustainable Future.* 10th ed. Upper Saddle River, N.J.: Benjamin Cummings/Pearson, 2010.

De Steiguer, J. E. *The Origins of Modern Environmental Thought.* Tucson: University of Arizona Press, 2006.

Dowie, Mark. *Losing Ground: American Environmentalism at the Close of the Twentieth Century.* Cambridge, Mass.: MIT Press, 1995.

Kline, Benjamin. *First Along the River: A Brief History of the United States Environmental Movement.* 3d ed. Lanham, Md.: Rowman & Littlefield, 2007.

O'Neill, John. *Ecology, Policy, and Politics: Human Well-Being and the Natural World.* New York: Routledge, 1993.

## Nature reserves

CATEGORY: Preservation and wilderness issues
DEFINITION: Areas set aside for the protection of animal and plant species, as well as unique geologic formations and other landscape features, in their natural state
SIGNIFICANCE: Protected natural areas serve as repositories of biodiversity. Their preservation helps protect complex ecological, hydrological, and climatological cycles, as well as distinctive landscapes and cultural and historical resources. However, the protection of these sites from development and the exploitation of the natural resources they contain often leads to conflict and controversy.

The nature reserve concept is often traced back to 1872, when the U.S. Congress established Yellowstone as the world's first national park. In fact, the basic idea is much older. In medieval England, the New Forest was a royal preserve. It had special protected status so that the king and his nobles could enjoy fine hunting in a land where farming was rapidly encroaching upon woodlands. Despite the hunting and poaching detailed in history and legend, the New Forest remained a viable natural area for centuries.

As the vast western regions of North America were explored, spectacular places such Banff in Canada (1885) and Yosemite in California (1890) received government protection. The number of protected areas grew over the following decades. By the late twentieth century, thousands of natural sites and parks had been designated for protection at state and federal levels in North America. These range from pristine wilderness to multiple-use areas such as national forests, where some commercial logging is allowed.

The movement to set aside areas for the preservation of nature developed differently on other continents, but by the late twentieth century almost every nation in the world had established nature reserves. By 1980 more than 3,000 protected sites existed worldwide; three decades later, the total exceeded 100,000 sites that cumulatively represented more than 12 percent of the planet's land surface. International treaties give special recognition to three types of reserves: Biosphere reserves, designated by the United Nations Educational, Scientific, and Cultural Organization (UNESCO), shield strictly protected areas with zones of limited, sustainable activity around them; World Heritage Sites, also designated by UNESCO, contain extraordinary natural, cultural, or historical features; and Ramsar sites are wetlands systems protected under the Convention on Wetlands of International Importance, also known as the Ramsar Convention. Most nature reserves are created and managed by governments, although some, often referred to as nature conservancies, are run under private auspices.

Nature reserves serve as reservoirs of native species and genetic diversity; as viable ecosystems that support biological, hydrological, and climatic cycles even outside their bounds; and as places for people to observe and enjoy the natural world. These goals are not innately incompatible. In managing reserves, however, decisions may have to be made that favor one aspect over the others.

Issues in Nature Reserve Management

Few people oppose nature reserves in the abstract, but when it comes to real-life choices, many conflicts arise, and clashes between biological or scenic integrity and human economic needs are common. In developed countries, disagreements often take the form of duels between industries seeking to use specific natural resources and government bureaus and conservation groups trying to prevent such activities. Examples include disputes over oil drilling or pipeline construction in remote natural areas and clashes over lumbering and mining in national forests. Such struggles become more heated when unique species are threatened, as in the controversy in the Pacific Northwest regarding logging in the forest habitat of the northern spotted owl.

Developing nations face similar problems and more. By restricting natural areas, poorer countries often deprive their own citizens of traditional food sources and trade items. Banning the sale of natural products may encourage poaching, as the banned products may bring even higher prices on the black market. A well-known example is the illegal trade in African elephant ivory and rhinoceros horns.

Tensions also occur between individuals and groups who value the preservation of pristine wilderness and those who demand access to natural areas for recreational use. Tourists and campers bring cash and public support to parks, but they can disturb delicate natural balances. Automobile traffic brings pollution and noise problems inside the boundaries of nature reserves. In spite of strong warnings about behavior, visitors may start forest fires or introduce alien organisms to an area. Advocates of public access argue, however, that because such lands are public trusts, their reasonable use should not be denied to the citizens who support them.

Other issues arise concerning the management of nature reserves. Fires, epidemics, and unusual weather patterns may threaten nature's balances. When disasters occur, should humans intervene? A consistent environmental ethic might say no—nature will eventually restore itself. However, entire species may be lost before a natural balance reasserts itself. Many such losses to biodiversity are at least partly or indirectly caused by human actions. Science does not provide clear answers. Furthermore, measures taken to prevent or relieve such crises do not always work and may have unanticipated consequences of their own. A number of technical issues related to nature reserves also require further consideration. These include examination of the efficacy of small reserves in protecting specific species, the role of reserves in stabilizing the mix of gases in the earth's atmosphere and slowing global climate change, and the tipping point in the recovery of a species or a damaged ecosystem.

Several promising ideas have enriched the reserve movement. The foremost is international cooperation and joint action. UNESCO's World Heritage Sites program is a major achievement of this approach. Another is the establishment or recognition of natural areas sponsored by nongovernmental groups. Two different but equally positive examples in the United States are the nature conservancy movement and the increasing recognition of the importance of Native American sacred sites that began in the late twentieth century. Yet another concept that has helped reserve efforts to flourish, particularly in the developing world, is ecotourism. Ecotourist dollars represent an economic incentive for countries to preserve their unspoiled wilderness areas.

*Emily Alward*
*Updated by Karen N. Kähler*

FURTHER READING

Allin, Craig W., ed. *International Handbook of National Parks and Nature Reserves*. New York: Greenwood Press, 1990.

Carey, Christine, Nigel Dudley, and Sue Stolton. *Squandering Paradise? The Importance and Vulnerability of the World's Protected Areas*. Gland, Switzerland: World Wide Fund for Nature International, 2000.

Chape, Stuart, Mark D. Spalding, and Martin D. Jenkins, eds. *The World's Protected Areas: Status, Values, and Prospects in the Twenty-first Century*. Berkeley: University of California Press, 2008.

Duncan, Dayton, and Ken Burns. *The National Parks: America's Best Idea—An Illustrated History*. New York: Alfred A. Knopf, 2009.

Ghimire, K. B., and Michel P. Pimbert, eds. *Social Change and Conservation*. Sterling, Va.: Earthscan, 2009.

Lockwood, Michael, Graeme Worboys, and Ashish Kothari, eds. *Managing Protected Areas: A Global Guide*. Sterling, Va.: Earthscan, 2006.

Riley, Laura, and William Riley. *Nature's Strongholds: The World's Great Wildlife Reserves*. Princeton, N.J.: Princeton University Press, 2005.

# Preservation

CATEGORIES: Preservation and wilderness issues; philosophy and ethics

DEFINITION: Maintenance of wilderness areas in an undisturbed state

SIGNIFICANCE: While debates continue regarding the best uses of public lands, it is clear that proponents of preservation have played a large role in preventing the conversion of many wilderness areas to development, logging, mining, or other uses.

The concept of preservation of wilderness emerged in the United States in the nineteenth century as a response to the large-scale disposal of public lands then taking place and to such economic activities as mining and logging, which had altered much of the western landscape. Preservation is typically contrasted with conservation, which allows managed exploitation of resources for economic purposes. John Muir, who is usually cited as the first American preservationist, condemned the common perception of wilderness as an economic resource. Muir and other preservationists argued that the American wilderness possesses an inherent value that must be protected from commercial exploitation. Most historians point to the battle over the damming of Yosemite's Hetch Hetchy Valley at the beginning of the twentieth century as the first major conflict between conservationists and preservationists.

Although preservationists successfully excluded most commercial development from the national park system, conservation remained the dominant ethic in land-use management in the United States for the first half of the twentieth century. However, preservation gained a powerful new constituency in the years after World War II. A rising standard of living allowed growing numbers of Americans the opportunity to pursue leisure activities. Many such Americans, who typically lived in urban or suburban areas, supported preservation because it afforded them unspoiled settings in which to enjoy outdoor activities such as camping.

The growing demand for wilderness areas influenced government officials charged with setting land-use policy, culminating in 1964 with the passage of the Wilderness Act. Throughout the 1960's and 1970's, politicians responded to the American public's continued support for preservation. However, during the 1980's and into the 1990's, government officials increasingly favored multiple-use management of public lands over preservation.

The history of preservation shows that it served as a response to the success of capitalism as an economic system. Ironically, resource exploitation allowed Americans the income and free time to enjoy the outdoors while at the same time significantly altering the environment. For many Americans, the negative consequences of the exploitation of natural resources began to outweigh the benefits that development brought to society. Environmentalists articulated this concern, which even appeared in the wilderness legislation of the 1960's. Politicians, including President Lyndon B. Johnson, began discussing the importance of beauty and the value of nature.

Critics have contended that the preservation ethic is an expression of middle- and upper-class Americans who already enjoy the benefits of economic exploitation of resources; the needs of lower-class citizens who would profit from the continued development of natural resources are disregarded in favor of leisure activities for the affluent. However, preservation constitutes more than an approach to land-use management that calls for the maintenance of natural playgrounds for middle-class Americans. It also serves as more than a muddled critique of capitalism. It expresses a value system, and as such it could be counted as a philosophy that has important political and social implications.

Preservation Versus Conservation

In discussions of preservation's philosophical import, preservation is usually contrasted with conservation. Whereas conservation understands nature in terms of resources that have value to human economies, preservation regards nature as possessing additional value to humans, bringing aesthetic and even spiritual qualities to human life. Conservationists perceive forests, minerals, and wildlife as separate categories, whereas preservationists argue that such categorization is artificial. Preservationists assert that nature must be understood as a whole in which the apparently individual parts are intimately interconnected in ways that humans, who are a part of the natural system, cannot fully understand. The failure to preserve wilderness thus has consequences that ripple through the entire ecosystem, including human society.

Some environmentalists argue that the distinction between conservation and preservation, while useful for defining advocacy groups and management plans, is simplistic and misleading. Rather than perceiving the two as mutually exclusive, adversarial positions, they maintain that preservation and conservation are two tools for understanding the human relationship to the environment. Some radical environmentalists dismiss preservation on the grounds that, like conservation, it assesses nature according to human values, albeit a wider range of such values than are applied by conservation. They maintain that preservation ultimately fails as a philosophy because it does not understand that nature exists wholly apart from any value system. It is thus faulty because it does not lead to the complete reform of human perceptions necessary to end the destruction of the environment.

Despite such criticisms, preservation remains an important environmental ethic at the beginning of the twenty-first century. In the United States and elsewhere, preservationists can claim that their efforts have saved wilderness regions that otherwise would have been devastated by mining, logging, or other activities. These victories, however, also require preservationists to address difficult issues regarding the management of wilderness areas; lively debates continue around such issues as scientific study within wilderness preserves, fire suppression, and the reintroduction of wildlife species, such as wolves. Moreover, preservationists often must work to ensure the establishment of standards concerning air and water quality outside preserved areas in order to protect the integrity of the wilderness.

*Thomas Clarkin*

Further Reading

Allin, Craig W. *The Politics of Wilderness Preservation.* 1982. Reprint. Fairbanks: University of Alaska Press, 2008.

Cawley, R. McGreggor. *Federal Land, Western Anger: The Sagebrush Rebellion and Environmental Politics.* Lawrence: University Press of Kansas, 1993.

Davis, Charles, ed. *Western Public Lands and Environmental Politics.* 2d ed. Boulder, Colo.: Westview Press, 2001.

Lewis, Michael. *American Wilderness: A New History.* New York: Oxford University Press, 2007.

Nash, Roderick. *Wilderness and the American Mind.* 4th ed. New Haven, Conn.: Yale University Press, 2001.

Oelschlaeger, Max. *The Idea of Wilderness: From Prehistory to the Age of Ecology.* New Haven, Conn.: Yale University Press, 1991.

# Ramsar Convention on Wetlands of International Importance

Categories: Treaties, laws, and court cases; preservation and wilderness issues
The Convention: International agreement that provides a framework for the conservation and wise use of wetlands
Date: Opened for signature on February 2, 1971
Significance: As the first international agreement to address issues surrounding the conservation of natural resources, the Ramsar Convention broke new ground in global environmental efforts.

During the 1960's concerns began to grow regarding the decline in populations of waterfowl in many parts of the world. One of the major factors causing this trend was the reduction in the number and size of the world's wetlands, which are habitats heavily used by waterfowl. In order to address these in-

*A great blue heron on a jetty in Chesapeake Bay. The Chesapeake Bay estuarine complex is one of the Ramsar Convention's named wetlands of international importance.* (©Dreamstime.com)

terrelated environmental issues, representatives of various nations came together to develop the Convention on Wetlands of International Importance. This agreement, reached by delegates from eighteen nations on February 2, 1971, in Ramsar, Iran, broke new ground in global environmental efforts; it was the first international treaty to address the conservation and wise use of a natural resource.

The Ramsar Convention came about because the importance of wetlands to the global environment and the needs of humans, as well as the need for an international approach to deal with wetlands, had become apparent. Wetlands are important because of their high biodiversity and their role in water purification, water storage, flood abatement, and groundwater recharge. Many wetlands extend across national boundaries; for example, fish may hatch in the wetlands of one country but be caught as adults in those of another country. Also, many birds migrate hundreds or thousands of kilometers twice each year and need to rest, feed, and breed in the wetlands of many countries.

The Ramsar Convention, which entered into force on December 21, 1975, has been ratified by more than 150 nations, and the list of areas designated as "wetlands of international importance" has grown to more than 1,800. Key issues of the convention include urging member nations to develop management plans for wetlands that include the involvement of local communities and indigenous peoples, to promote wetlands education, and to establish monitoring programs that have the ability to detect changes in the ecological character of wetlands. Under the treaty, each country is obligated to implement the convention in four basic ways: It must designate at least one wetland for inclusion in the list of wetlands of international importance, it must include wetlands conservation and wise use as a major focus within its national land-use planning, it must promote wetlands conservation by establishing nature reserves on wetlands and promoting wetlands education, and it must consult with other countries concerning the implementation of the convention.

The policy-making body of the convention is the Conference of the Contracting Parties. Each member nation sends representatives to a conference every three years for the purpose of receiving and reviewing reports on the work of the convention and for approving the work and budget of the convention for the next three years. The Ramsar Convention is administered by the Ramsar Bureau, located in Gland, Switzerland. The bureau is advised on a regular basis by the Standing Committee and the Scientific and Technical Review Panel, each of which is composed of representatives from member nations. The work of the convention is funded by contributions from those nations.

*Roy Darville*

FURTHER READING

Hunt, Constance Elizabeth. *Thirsty Planet: Strategies for Sustainable Water Management.* New York: Zed Books, 2004.

Smardon, Richard C. *Sustaining the World's Wetlands: Setting Policy and Resolving Conflicts.* New York: Springer, 2009.

# Recycling

CATEGORY: Waste and waste management

DEFINITION: Processing of used materials into fresh supplies of the same materials for new products or salvaging of certain materials from more complex products

SIGNIFICANCE: Recycling constitutes the primary component of modern waste management. It benefits the environment by reducing waste, conserving resources, saving energy, and reducing air and water pollution that can result from waste disposal methods, thus protecting human health. In addition, recycling can provide fiscal benefits, such as by expanding manufacturing jobs.

In 2009 the United States produced 251 million tons of trash, with 82 million tons (32.5 percent) of those materials recycled. In the first decade of the twenty-first century, total recycling in the United States increased approximately 100 percent. At the national level, the U.S. Environmental Protection Agency (EPA) oversees waste management, including hazardous wastes and landfills, and sets recycling goals. However, no national laws for recycling exist; instead, individual state and local governments have created their own laws concerning recycling.

Several U.S. states have laws establishing deposits and refunds for beverage containers. Other states ban the deposition of recyclable materials into landfills. Some cities, including New York and Seattle, have passed laws that include fines for failing to recycle certain materials. Several organized voluntary and educational programs have been established to increase recycling where it has not been mandated by law. Recycling education is usually integrated into science or social studies classes at the elementary, middle school, and high school levels. November 15 is celebrated as America Recycles Day, which is dedicated to raising awareness about the importance of recycling and to encouraging Americans to recycle and buy recycled products.

For recycling to be economically feasible, efficiently managed, and environmentally effective, adequate recyclable materials must be available, a system must be put in place to extract those materials from the waste stream, and a facility must be available locally where the materials can be reprocessed. In addition, there must be demand for the recycled products.

Recyclable materials are generally collected in three ways: through curbside pickup at consumer homes and businesses, through consumer delivery to drop-off centers, and through consumer return through deposit or refund programs. After collection, recyclables are sent to materials recovery facilities, where they are sorted and prepared for processing into marketable items made entirely or partially of recycled content.

In 2008 the United States generated some 250 million tons of municipal solid waste (MSW), commonly called trash or garbage. Organic materials such as yard trimmings, paper products, and food wastes make up more than two-thirds of human trash. This organic waste could become useful compost instead of being deposited in landfills. In composting, natural aerobic bacteria break down organic materials into fertile topsoil for plant cultivation. Humus and its nutrients in compost help regenerate and enrich soils, help remediate contaminated soils, prevent pollution, and provide economic benefits by, for example, reducing the need for fertilizers and pesticides.

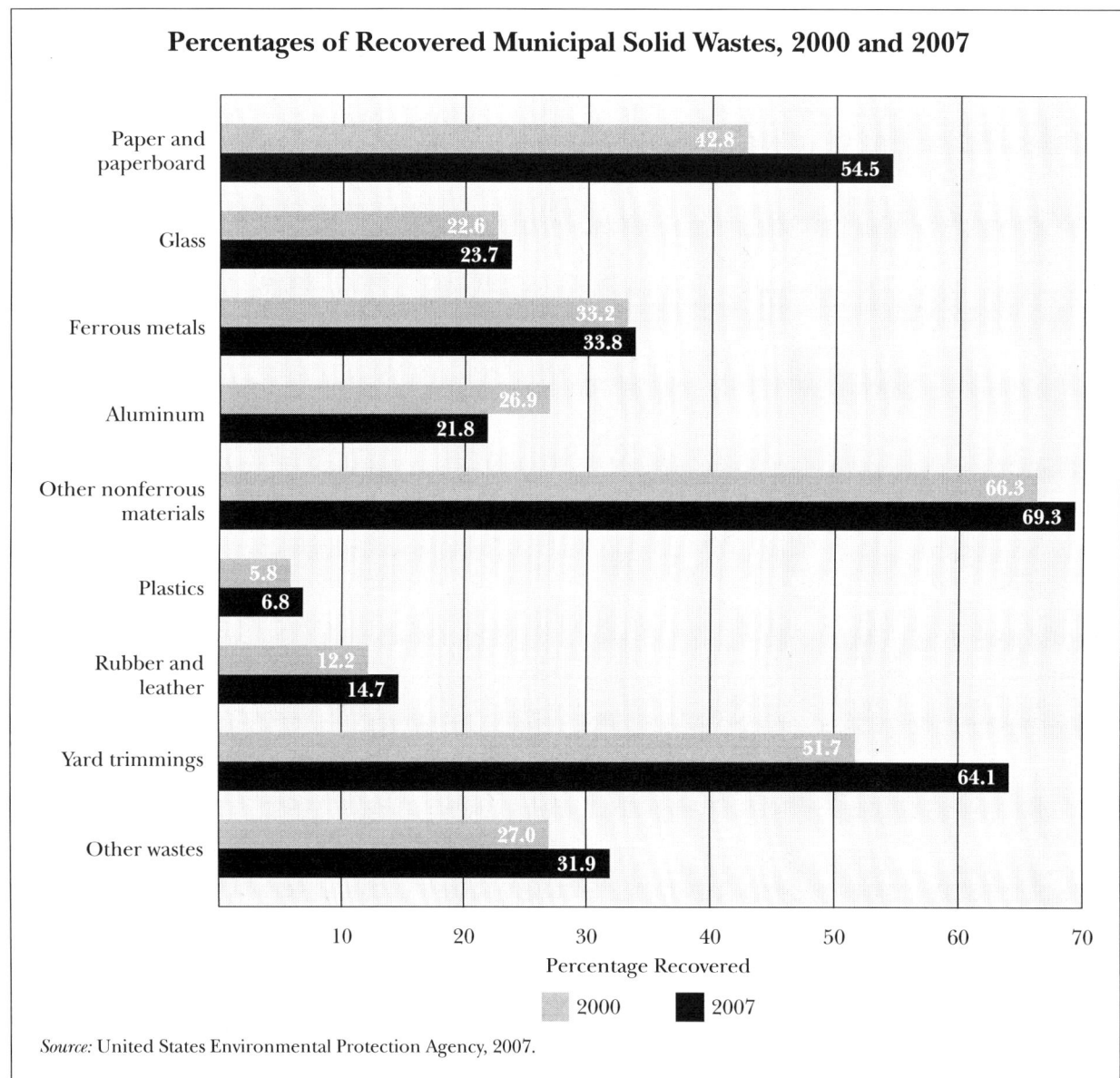

*Source:* United States Environmental Protection Agency, 2007.

## COMMONLY RECYCLED MATERIALS

Paper is the most common material in municipal solid waste, approximately 35 percent of the total. Americans recycled more than 50 percent of paper used in 2008, but that percentage could be increased. Approximately 80 percent of the paper mills in the United States are designed to depend on paper recycling; they reduce recycled paper to pulp, which is then combined with pulp from newly harvested wood. Wood fiber can be recycled only up to five times, however, since damage to the fibers with each recycling process decreases the quality of the resulting product; hence, more new fibers must be added with each cycle.

Recycled fabric items are sorted into grades; some are then processed to make industrial wiping cloths, whereas others are made into filling products or used in the manufacture of paper. Discarded clothing items often end up in landfills in the United States, but many unwanted but still wearable clothes can be recycled through charitable giving. This practice helps to reduce unwanted waste and conserves the resources needed to make new clothing, while providing clothing to persons in need.

Plastic recyclable items are sorted and sent to facilities where they are washed and ground into small flakes that are dried, melted, filtered, and formed into pellets used to manufacture new plastic products. In 2008 the U.S. MSW stream included some 13 million tons of plastics—about 12 percent, an increase from less than 1 percent in 1960. The recycling of plastics has been found to reduce energy use in the United States by 26 percent. Given that 70 percent of the plastics made in the United States are made by factories that use domestic natural gas, the energy saved is freed up for other uses, such as heating and cooling homes.

When glass bottles and jars are recycled, the containers are first smashed and the broken glass, or cullet, is examined for purity and cleaned of contaminants at a glass recycling plant. The cullet is then further crushed and placed in a melting furnace with a mix of raw materials. The resulting material is then mechanically blown or molded into new bottles, jars, and other glass items. Of all the glass recycled in the United States, 90 percent goes into the making of new containers. Glass cullet is also used for aggregate and "glassphalt" for road construction (glassphalt contains 30 percent recycled glass). Unlike paper, glass can be recycled over and over again, as reprocessing does not affect glass's structure. Because cullet costs less than the raw materials needed to make new glass, recycling glass saves manufacturers money at the same time it conserves natural resources; it also helps glass-making furnaces last longer because it melts at a relatively low temperature. Of the 12.2 million tons of glass that were part of the U.S. MSW stream in 2008, 2.8 million tons, or 23 percent, was recovered for recycling. That figure represents a significant increase from 750,000 tons in 1980.

Iron and steel, the world's most commonly recycled ferrous metals, are generally separated from the waste stream through the use of magnets. Recycled steel is processed at steelworks, where scrap steel is remelted and forged into new products. Like glass, steel can be recycled repeatedly with no reduction in quality.

Aluminum is one of the most widely recycled nonferrous metals and one of the most efficiently processed. Approximately 40 percent of aluminum in the average aluminum can is recycled material. Aluminum beverage containers are the largest source of aluminum in the U.S. MSW stream; in 2008 they accounted for 3.4 million tons (1.3 percent) of MSW, up from only 0.4 percent in 1960. The aluminum recycling process uses 92 percent less energy than is needed to produce aluminum from bauxite ore, in part because the temperature necessary to melt recycled aluminum is 600 degrees Celsius (1,112 degrees Fahrenheit), in contrast to the 900 degrees Celsius (1,652 degrees Fahrenheit) needed to extract mined aluminum from bauxite. The processes of recycling and reuse produce no changes in aluminum, so it can be recycled repeatedly.

Electronic wastes pose problems of energy cost and release of toxic substances that can be reduced by recycling. In many parts of the world the disposal of computers, televisions, mobile phones, and other electronic devices into the general waste stream is forbidden because of the toxic contents of certain components in these products. In the United States, the EPA advises consumers and local governments regarding the safe recovery and recycling of components of electronic waste. Electronic waste recycling plants process discarded electronic items by taking them apart and separating the various metal, plastic, and other components, collecting those that can be reused.

Batteries are sometimes recycled, but because of wide variations in sizes and types, recycling is not always practical. Those that cannot be recycled must be disposed of carefully, as batteries contain heavy metals that can pollute soil and water. The lead-acid batteries used in motor vehicles are easily recycled, and in the United States many local and state governments require businesses that sell such batteries to receive and recycle used batteries from consumers. The recycling rate for automotive batteries in the United States is 90 percent; new batteries generally contain up to 80 percent recycled material.

Tires are frequently recycled at an industrial level. Used tires are ground up and the resulting material is used for a variety of purposes, including insulation and road surfacing. Rubber mulch made from old tires is used to soften the surfaces of playgrounds. Tires are also sometimes recycled into consumer products. Recycled rubber from tires is used to make items such as doormats, trash cans, and even messenger bags; also, a number of companies make shoes that have soles made of recycled tire treads.

Technologies are advancing to produce improved materials for faster recycling, such as completely biodegradable packing materials. An example is a plant-based, all-natural material that breaks down into inert proteins when it comes into contact with water and is consumed by soil bacteria to produce a product used as lawn fertilizer. Scientists at Sony discovered that expanded polystyrene foam, such as that used as packing "peanuts" for cushioning fragile items inside boxes, completely dissolves at room temperature when sprayed with limonene (a natural oil extracted from the skins of citrus fruits) and can be processed for reuse.

### Debates

Some critics of the promotion of recycling as a solution to environmental problems have suggested that the modern waste management system has fundamental flaws that call for reexamination—they have therefore added a "fourth R," rethink, to the "three R's": reduce, reuse, and recycle. They argue that the costs of collecting and transporting recyclable materials for processing far outweigh the costs of production processes using raw materials. They also assert that more jobs are lost through the reduced collection of raw materials than are created by the recycling industry, which, additionally, pays low wages and offers poor working conditions. Critics argue further that when all processes are considered, the production of recycled products consumes more energy than would the traditional landfill disposal of the recycled materials used to make the products.

Recycling proponents counter that the benefits of recycling compensate for any higher monetary costs it may create. Landfilled wastes pollute groundwater and waterways and contribute significantly to global warming through the release of methane into the atmosphere, producing long-term financial costs of pollution remediation. Proponents also argue that the workers that would be needed to gather amounts of virgin materials equal to the amounts provided through recycling would be working in jobs, such as timber harvesting and ore mining, that have much more dangerous workplace conditions than are found in the recycling industry. Recycling proponents note also the reductions in energy needs represented by the recycling of such materials as paper and aluminum compared with the processing of raw materials. The EPA strongly supports recycling, emphasizing that by saving energy, recycling reduces emissions of carbon dioxide, a greenhouse gas linked with global warming.

*Samuel V. A. Kisseadoo*

### Further Reading

Ackerman, Frank. *Why Do We Recycle? Markets, Values, and Public Policy.* Washington, D.C.: Island Press, 1997.

Loeffe, Christian V., ed. *Trends in Conservation and Recycling of Resources.* Hauppauge, N.Y.: Nova Science, 2006.

McKinney, Michael L., Robert M. Schoch, and Logan Yonavjak. "Resource Use and Management." In *Environmental Science: Systems and Solutions.* 4th ed. Sudbury, Mass.: Jones and Bartlett, 2007.

Tammemagi, Hans. *The Waste Crisis: Landfills, Incinerators, and the Search for a Sustainable Future.* New York: Oxford University Press, 1999.

Weeks, Jennifer. "Future of Recycling: Is a Zero-Waste Society Achievable?" *CQ Researcher* 17, no. 44 (December 2007): 1033-1060.

Williams, Paul T. "Waste Recycling." In *Waste Treatment and Disposal.* 2d ed. Hoboken, N.J.: John Wiley & Sons, 2005.

## Roadless Area Conservation Rule

CATEGORIES: Treaties, laws, and court cases; preservation and wilderness issues

THE LAW: Federal policy initiative designed to protect national forests from commercial development

DATE: Issued on January 12, 2001

SIGNIFICANCE: To ensure the natural condition of the wilderness forests under federal jurisdiction and to maintain their integrity as complex ecosystems, Bill Clinton's presidential administration issued the Roadless Area Conservation Rule, a landmark federal environmental initiative, declaring part of the national forest preserves as "roadless," thus protected from any new mining, lumbering, drilling, and housing development interests.

Following World War II the U.S. Forest Service routinely approved private corporation projects aimed at extracting the reserves of natural resources—principally mining and logging interests—in otherwise protected national forests, a little more than 57 million hectares (140 million acres). During the mid-1990's, amid growing concerns over global warming and emerging scientific evidence that indicated the critical importance of protecting forests, Bill Clinton's presidential administration commissioned extensive public hearings into shaping a responsible plan to protect the national forests. Literally in its closing days, the administration issued the Roadless Area Conservation Rule, which essentially prohibited all development—logging, mining, drilling—on more than 23.5 million hectares (58 million acres) of forests and grasslands, roughly one-third of the national preserve, mostly in the West. Administration policy experts argued that the blanket protection would create secure habitat for wildlife, birds, and plant life and help protect the watershed supply for more than one-fifth of the country. In addition, it would provide for public recreation, hiking, kayaking, biking, and camping. The initiative permitted only emergency actions, such as clearing trees against a wildfire or cutting small-diameter trees only as necessary for habitat preservation.

Although environmentalists hailed the bold initiative, development interests and a phalanx of conservative western activists and politicians objected to the initiative as an unconstitutional overreach of federal power that would put at risk state economies. Quickly, the incoming presidential administration of George W. Bush delayed the initiative, calling for additional investigation into its scope. The Bush administration sought to give states authority to designate areas as roadless. In short order, a number of western states, most prominently Alaska, filed suit against the Clinton rule. After four years of contentious litigation, the Bush administration repealed the rule in May, 2005, and announced it would base decisions for setting aside public lands on state-by-state petitions to the Forest Service. That rule was quickly challenged in federal district court by several states and environmental activists (spearheaded by the environment-focused law firm Earthjustice). In September, 2006, the courts ruled the Bush protocol illegal, saying that a state-by-state plan violated the mandates of the National Environmental Policy Act of 1969 and the Endangered Species Act of 1973. The original Clinton rule was reinstated. Shortly after, contracts for oil and gas projects (more than three hundred) approved by the Forest Service during the years when the rule had been suspended were summarily voided. In August, 2008, however, a Wyoming federal district court once again invalidated the original Roadless Rule.

Seeking a plan that would mediate the controversy, and acting under pressure from both environmental activists and congressional representatives concerned about the open expansion by development interests, President Barack Obama's administration, while filing an appeal of the Wyoming decision, moved in May, 2009, to place decisions on a case-by-case basis solely with the secretary of agriculture for one year. That directive was renewed for an additional year in May, 2010, as the administration, confronting the catastrophic BP *Deepwater Horizon* oil spill in the Gulf of Mexico, began shaping a comprehensive environmental policy initiative.

*Joseph Dewey*

FURTHER READING

Bevington, Douglas. *The Rebirth of Environmentalism: Grassroots Activism from the Spotted Owl to the Polar Bear.* Washington, D.C.: Island Press, 2009.

Turner, Tom. *Roadless Rules: The Struggle for the Last Wild Forests.* Washington, D.C.: Island Press, 2009.

Vaughan, Jacqueline, and Cortner, Hannah J. *George W. Bush's Healthy Forests: Reframing the Environmental Debate.* Boulder: University Press of Colorado, 2005.

## Roosevelt, Theodore

CATEGORIES: Activism and advocacy; preservation and wilderness issues
IDENTIFICATION: American politician and conservationist who served as governor of New York and president of the United States
BORN: October 27, 1858; New York, New York
DIED: January 6, 1919; Oyster Bay, New York
SIGNIFICANCE: During his years as U.S. president, 1901-1909, Roosevelt did more to boost conservation efforts in the United States than any other president before him.

Born into an aristocratic family in New York City, Theodore Roosevelt was a sickly child. In his youth he collected animals both dead and alive, and he considered becoming a biologist while a student at Harvard. During the 1880's, after his first wife's death, Roosevelt sought consolation in the Dakota Badlands as a hunter and a rancher. Hunting was a passion with Roosevelt, but it was combined with a love of nature and considerable scientific knowledge about birds, animals, and plants. In 1887 he was a founding member and the first president of the Boone and Crockett Club, which, while devoted to hunting, also became

*Theodore Roosevelt, left, with naturalist and preservationist John Muir at Glacier Point above Yosemite Valley.* (Library of Congress)

one of America's earliest conservation organizations.

If the outdoors was Roosevelt's avocation, politics was his vocation. He held a number of political positions and became a national figure while leading the famous Rough Riders cavalry regiment during the Spanish-American War of 1898. After serving as governor of New York, he was elected vice president of the United States in 1900; when President William McKinley was assassinated in September, 1901, Roosevelt became the youngest president in U.S. history up to that time.

As president, Roosevelt made momentous strides in the field of conservation. In fact, the word "conservation" in the sense it is understood today came into use only during his presidency. It was a crucial time, for many of the nation's natural resources that had seemed inexhaustible to Thomas Jefferson one century earlier had been greatly depleted. Roosevelt had many friends who were equally devoted to nature, and while he was president he camped with John Burroughs in Yellowstone and John Muir in Yosemite. Unlike Muir, however, Roosevelt was not an environmental preservationist. Roosevelt agreed that much of the wild should be preserved in its natural state, but he and Gifford Pinchot, his chief forester and major conservation adviser, also believed that conservation means wisely using nature's resources to benefit later generations.

Promoting water reclamation in the West through the building of dams and irrigation projects was among Roosevelt's successful presidential acts, as was the establishment of millions of hectares of forest reserves (over the opposition of many in Congress) and the creation of fifty-one wildlife refuges. Five new national parks were added during Roosevelt's time in office, including Colorado's Mesa Verde, Oregon's Crater Lake, and South Dakota's Wind Cave Park. Through the Antiquities (or National Monuments) Act of 1906, Roosevelt created eighteen national monuments, including Devils Tower in Wyoming, the Petrified Forest and Grand Canyon in Arizona, and Muir Woods in California.

In 1908, toward the end of his second term as president, Roosevelt organized a three-day White House conservation conference for the nation's governors. This unprecedented event not only gave considerable publicity to the conservation cause but also resulted in the establishment of conservation commissions in thirty-six U.S. states. In February, 1909, Roosevelt hosted the North American Conservation Conference and proposed an international conservation conference, but that dream died when he left the presidency.

*Eugene Larson*

FURTHER READING

Brinkley, Douglas. *The Wilderness Warrior: Theodore Roosevelt and the Crusade for America.* New York: HarperCollins, 2009.

Morris, Edmund. *Theodore Rex.* New York: Random House, 2001.

# Rowell, Galen

CATEGORY: Preservation and wilderness issues
IDENTIFICATION: American nature photographer
BORN: August 23, 1940; Oakland, California
DIED: August 11, 2002; Bishop, California
SIGNIFICANCE: One of the best-known American nature photographers, Rowell was also a serious mountaineer and helped to develop the field of participatory photography.

Born in Oakland, California, Galen Rowell grew up in nearby Berkeley, where his father was a professor at the University of California and his mother was a concert cellist. Rowell was a good student but underperformed for his level of ability. He was curious and very bright. His family's devotion to the outdoors, and particularly California's Sierra Nevada mountain range had a profound influence on his life. From infancy he spent a few weeks of every summer camping in the Sierras, where he developed a love for the beauty of the mountains and the wilderness experience. He was also influenced by the rock-climbing section of the Sierra Club and was himself soon climbing mountains. While undertaking increasingly challenging mountaineering expeditions, he became an accomplished photographer.

Like his hero, photographer Ansel Adams, Rowell believed light was everything in a picture, and, like Adams, he was very patient and would go way out of his way to take a photograph at the moment when the lighting was perfect. In contrast to Adams, who worked only in black and white, using huge cameras with big negatives, Rowell used color and shot in 35mm most of his career. Whereas Adams's pictures were usually staged and shot from a tripod, many of Rowell's were close-ups, and many were taken as the photographer stood on a cliff or dangled from a rope. Rowell developed into one of the world's foremost mountain photographers. Although he is particularly known for his work in Yosemite and the High Sierra, he also produced highly respected photographs of the Himalayas and many other mountain ranges all over the world.

Having dropped out of college to work in his own small automobile shop to support his climbing and photographic interests, Rowell finally began to pursue his chosen career full time in 1972. His fame exploded in 1973 and 1974, when *National Geographic* magazine commissioned him to do a feature photographic article on climbing Yosemite's Half Dome. He recruited two other excellent rock climbers who had photography skills, and in 1973 they made the first "clean climb" of Half Dome—that is, they ascended the sheer monolithic cliff using no pitons. They trusted their lives to their hands, feet, and ability, using small metal wedges that they placed in existing cracks in the granite to support their ropes.

With the success of the *National Geographic* article, Rowell began to be sought out by other magazines, various societies, and individuals to lead groups and produce books. An articulate writer as well as a talented photographer, he eventually published more than twenty books and numerous articles and book chapters. His books were beautifully illustrated with his own photographs of mountain scenery. Perhaps his most highly regarded book is *Mountain Light: In Search of the Dynamic Landscape* (1986); other notable books include *In the Throne Room of the Gods* (1977) and an edition of John Muir's *The Yosemite* illustrated by Rowell's photographs (2001).

In 1984 Rowell received the Ansel Adams Award for his contributions to the art of wilderness photography. By that time he was devoting much of his time to teaching outdoor photography to small groups of both professional and amateur photographers on guided field trips. He was noted for his enthusiasm and patience with the amateurs. Rowell's life was cut short when he and his wife died in a small-plane crash in 2002.

*C. Mervyn Rasmussen*

### FURTHER READING

Rowell, Galen. *High and Wild: Essays and Photographs on Wilderness Adventure.* Bishop, Calif.: Spotted Dog Press, 2002.

_____. *Mountain Light: In Search of the Dynamic Landscape.* 2d ed. San Francisco: Sierra Club Books, 1995.

## Scenic Hudson Preservation Conference v. Federal Power Commission

CATEGORIES: Treaties, laws, and court cases; preservation and wilderness issues

THE CASE: U.S. Court of Appeals decision regarding basic principles of standing in U.S. environmental law

DATE: Decided on December 29, 1965

SIGNIFICANCE: The *Scenic Hudson* decision was a declaration of judicial authority that granted environmental groups with noneconomic interests the right to intervene in federal regulatory agency licensing decisions and held a federal regulatory agency responsible for engaging in its own evidentiary inquiry, considering alternatives to granting a license, and weighing the environmental impacts of its decisions.

In the early 1960's the Scenic Hudson Preservation Conference, a coalition of conservation and environmental groups, and the towns near Storm King Mountain in New York challenged the license that Consolidated Edison Company (Con Ed) had obtained to construct a storage reservoir, a powerhouse, and transmission lines on the Hudson River in the region. However, the Federal Power Commission (FPC) denied their motions to intervene and reopen the hearings to consider evidence on alternatives to the Storm King facility, the cost and practicality of underground transmission lines, and the feasibility of fish protection devices.

In the case *Scenic Hudson Preservation Conference v. Federal Power Commission*, the U.S. Court of Appeals dismissed the FPC's argument that Scenic Hudson lacked standing to seek judicial review of Con Ed's license because the organization had made no claim of personal economic injury. Scenic Hudson clearly had economic interests to protect. The Storm King reservoir would inundate 27 kilometers (17 miles) of hiking trails, and the planned aboveground transmission lines would decrease the value of public land and the tax revenues of private land and would interfere with community planning. Scenic Hudson also had noneconomic interests that would be affected by the Storm King project that were not barred from judicial review. The Court of Appeals held that the U.S. Constitution does not require an aggrieved party to have a personal economic interest, nor does the Federal Power Act (the FPC governing legislation), which permits "any aggrieved party" to obtain judicial review of an FPC order.

In terms of the licensing decision, the FPC had acknowledged that the "principal issue ... is whether the project's effects on the scenic, historical and recreational values are such that we should deny the application." Since the FPC's standing argument was at odds with its recognition of the importance of noneconomic issues in its licensing decision, the court broadly read the statute to grant standing to organizations, such as Scenic Hudson, that had special interests in "aesthetic, conservational, and recreational" issues to act as private attorneys general to ensure that the FPC would protect the public interest. Since Scenic Hudson was an aggrieved party, the court read the statute's language, which gave the FPC the discretion to admit as a party "any person whose participation in the proceeding may be in the public interest," to create "an absolute right of intervention."

After the appellate court settled the standing issue, it found that the Federal Power Act gives the FPC a specific planning responsibility for the development of waterways for commerce, water power, and recreation. The statute's planning responsibility requires the agency to weigh these three factors in making a licensing decision. Then the court broadly read the FPC's responsibility to consider recreational objectives to include "the preservation of natural resources, the maintenance of natural beauty, and the preservation of historic sights." In fact, the FPC regulations require a recreation plan.

The FPC, in granting Con Ed a license for the Storm King project, did not fulfill its statutorily mandated planning duties because it failed to develop a complete record. First, the FPC did not explore alternatives to the Storm King project, including the use of gas turbines or the purchase of power from interconnections with other electric utilities, nor did the agency require Con Ed to supply this information. Therefore, the agency was unable to determine whether Con Ed needed Storm King to meet its future power needs. Second, the FPC did not seriously consider the cost of underground transmission routes; it simply accepted Con Ed's estimates of their cost even though the estimates had been questioned by the FPC's staff and hearing examiner. As a consequence, the agency was unable to weigh "the aesthetic advantages of underground transmission lines against the economic disadvantages." Third, there was sufficient evidence before the FPC on the danger of the Storm King project to fish that the FPC should have made further inquiries, but the agency failed to do so. It even dismissed as "untimely" the petitions of several groups to intervene and present evidence on Storm King's destructive impact on the spawning grounds for striped bass and shad and the unavailability of any powerhouse screening devices to protect fish eggs and larvae.

The Court of Appeals set aside the FPC's licensing order because the record had failed to support the agency's decision and then remanded the case with instructions to reexamine the alternatives to Storm King, the cost of underground transmission lines, and the fish spawning issue. In sum, the Scenic Hudson opinion was a bold declaration of judicial authority that granted environmental groups with noneconomic interests the right to intervene in federal regulatory agency licensing decisions and held a federal regulatory agency responsible for engaging in its own evidentiary inquiry, considering alternatives to granting a license, and weighing the environmental impacts of its decisions.

The FPC subsequently held hearings and issued Con Ed a license to build the Storm King project, which the federal Court of Appeals affirmed. However, the Scenic Hudson case was only the beginning of a two-decade controversy over the Storm King project. Further litigation on the project's threat to fish life expanded to include six other power stations along the Hudson River once the newly created Environmental Protection Agency conducted extensive hearings under the Federal Water Pollution Control Act amendments of 1972. The controversy was not resolved until 1981, when seventeen months of mediation led to Con Ed's agreement to forfeit its Storm King license and donate the site as a public park.

*William Crawford Green*

FURTHER READING

Dunwell, Frances F. "The 1960s: Scenic Hudson, Riverkeeper, Clearwater, and the Nature Conservancy Campaign to Save a Mountain and Revive a 'Dead' River." In *The Hudson: America's River.* New York: Columbia University Press, 2008.

Hoban, Thomas, and Richard Brooks. *Green Justice: The Environment and the Courts.* 2d ed. Boulder, Colo: Westview Press, 1996.

Houck, Oliver A. "Storm King." In *Taking Back Eden: Eight Environmental Cases That Changed the World.* Washington, D.C.: Island Press, 2010.

Lewis, Tom. *The Hudson: A History.* New Haven, Conn.: Yale University Press, 2005.

Talbot, Allan. *Power Along the Hudson: The Storm King Case and the Birth of Environmentalism.* New York: Dutton, 1972.

## Serengeti National Park

CATEGORIES: Places; preservation and wilderness issues

IDENTIFICATION: Large national safari park in northern Tanzania, East Africa

SIGNIFICANCE: The Serengeti National Park provides protection to vast numbers of animal, insect, and plant species. In addition to being a popular tourist destination, the park is active in conservation and research efforts.

The name Serengeti, meaning "endless plain," derives from the language of the Masai, a seminomadic people native to East Africa who have grazed their cattle on the Serengeti Plain for more than two thousand years. The first European to see the area was Oskar Baumann, an Austrian explorer and cartographer, in 1892. Germany colonized the Serengeti until the end of World War I, when Tanganyika (now Tanzania) became a British protectorate. In 1921 the British designated 324 hectares (800 acres) of the plain as a partial game reserve to protect the lions; this area became a full reserve in 1929.

The British established the whole of the Serengeti as a national park in 1951, before Tanganyika achieved independence. The Masai tribespeople on the plain were relocated to the southeastern portion of the park, in the Ngorongoro highlands. In the 1950's a book on the Serengeti by Bernhard Grzimek, president of the Frankfurt Zoological Society, and a documentary film based on the book that Grzimek wrote and directed brought worldwide attention to the Serengeti National Park. In the early twenty-first century, the Frankfurt Zoological Society works closely with Tanzania's national park service, providing funding for research, training, and conservation at the Serengeti National Park.

The park is roughly 14,760 square kilometers (5,700 square miles) in area. To the north it borders Kenya, to the west Lake Victoria, and to the east the Great Rift Valley. On the south it is surrounded by the buffer zones of the Ngorongoro Conservation Area, set up in 1959, and the Masai Mara National Reserve in Kenya. Within the park, tourists are accommodated in five lodges, seven tented camps, and a number of camp sites. Administration of the park is centered in Seronera, a small settlement within its boundaries.

The topography of the park is widely varied. The southern and central areas consist of savanna interspersed with rocky outbreaks known as kopjes or koppies. To the west is a corridor of more fertile ground watered by the Grumeti River. To the north are the permanent watercourses of the River Mara in the Lobo area. One of the great sights in the park is the annual migration of millions of wildebeest, joined by some 200,000 zebras and 300,000 Thomson's gazelles, from south to north, a distance of some 966 kilometers (600 miles), and then their return. Such migrations are attended by many predators: lions, cheetahs, jackals, leopards, hyenas, and African wild dogs.

In addition to these animals, the park is home to

*Wildebeest cross the Mara River in Serengeti National Park during their annual migration.* (©Jannie Nikola Laursen/iStockphoto.com)

large herds of elephants, buffalo, impalas, giraffes, waterbucks, and other types of gazelles, as well as crocodiles that live in the rivers. The large population of black rhinoceros that was once present in the park has been decimated by poaching; the numbers of this species had become dangerously low by the early years of the twenty-first century. Also native to the Serengeti are some five hundred bird species, including the secretary bird, ostrich, and black eagle. At the insect level species are abundant as well. One hundred varieties of dung beetle have been identified, for example. After the rains, the ground in many parts of the park is carpeted with a wide variety of wildflowers. Owing to the great biodiversity found there, the Serengeti National Park has been designated a World Heritage Site by the United Nations Educational, Scientific, and Cultural Organization (UNESCO).

Environmental challenges for the park include the comparative fragility of its ecosystems for such large numbers of animals; drought and erosion have posed particular problems. In addition, some conflicts have arisen on the borders of the park because animals from the park have killed the livestock of local herdsmen; also, diseases are sometimes spread between the wild animals of the park and domesticated animals. Efforts to reduce such tensions have included the establishment of four Wildlife Management Areas that encompass some twenty-three villages near the park's borders.

*David Barratt*

FURTHER READING

Holmern, Tomas, Julius Nyahongo, and Eivin Røskaft. "Livestock Loss Caused by Predators Outside the Serengeti National Park, Tanzania." *Biological Conservation* 135, no. 4 (April, 2007): 518-526.

Homewood, K. W., and W. A. Rodgers. *Maasailand Ecology: Pastoralist Development and Wildlife Conservation in Ngorongoro, Tanzania.* 1991. Reprint. New York: Cambridge University Press, 2004.

Roodt, Veronica. *The Tourist Travel and Field Guide of the Serengeti National Park.* 4th ed. Hartebeesport, South Africa: Papyrus, 2005.

Turner, Myles. *My Serengeti Years: The Memoirs of an African Game Warden.* New York: W. W. Norton, 1987.

## Sierra Club

CATEGORIES: Organizations and agencies; activism and advocacy; preservation and wilderness issues

IDENTIFICATION: American environmental organization

DATE: Established in 1892

SIGNIFICANCE: The Sierra Club is one of the largest, oldest, and most influential conservation organizations in the world. Unique among large environmental groups because of its reliance on volunteer activists and its democratic structure, the Sierra Club pioneered many grassroots political techniques in its efforts to preserve wilderness and protect parks and other natural areas.

John Muir, amateur naturalist and writer, discovered for himself the wonders of Yosemite Valley in California's Sierra Nevada in 1868. He soon realized that such areas needed to be protected from development and resource extraction, and he used his writings to influence his friends and others to lobby for the creation of Yosemite National Park.

Muir conceived of the idea of an organization that would ensure the protection of Yosemite and the surrounding wildlands, and in 1892 the Sierra Club was born, with Muir as its first president. The club's stated purpose was

> to explore, enjoy, and render accessible the mountain regions of the Pacific Coast . . . and to enlist the support and cooperation of the people and government in preserving the forests and other natural features of the Sierra Nevada.

This statement set the tone for an organization that went on to combine the recreational goals of a hiking and climbing club with political savvy and influence. Muir's idea was to build a constituency for nature by getting people out into the mountains, showing them areas that needed to be saved, and explaining how those areas were endangered, a technique that the Sierra Club has now used successfully for more than a century.

After the relatively easy success of the creation of Yosemite National Park, the Sierra Club's next major campaign was to be more difficult and ultimately unsuccessful. In 1907 a dam and reservoir to supply water to San Francisco were proposed for Hetch Hetchy Valley, an area within Yosemite National Park that many considered to be nearly equal to Yosemite Valley in scenic grandeur. A long campaign to prevent the building of the dam, and thus the flooding of the valley, revealed dissent within the Sierra Club and made clear the difficulties inherent in the organization's strictly democratic structure. The battle was finally lost, and, through an act of Congress, the dam was built. Out of this failure, however, the Sierra Club gained experience in managing its own growing organization and building national support for its views by using media publicity and connections with a network of other conservation organizations.

Muir died in 1914, one year after the Hetch Hetchy defeat, but the Sierra Club continued to grow in membership and influence. The outings program introduced people to the wonders of the Sierra Nevada and other wild areas as far afield as Montana and the Canadian Rocky Mountains. Club members developed mountaineering and rock-climbing techniques, as well as low-impact camping ethics. On the conservation front, the club was involved in campaigns to create Kings Canyon and Olympic national parks and to prevent the damming of rivers in Yellowstone and Glacier parks.

In 1951 another proposed dam project pushed the Sierra Club to national prominence and influence. Another federally protected area, in this case Utah's Dinosaur National Monument, was to be the site for the Echo Park Dam and another dam as part of the Colorado River Project. Remembering the loss of Hetch Hetchy, the club vowed to fight harder this time and hired David Brower as its first executive director. Brower led the successful campaign against these dams with float trips on the river for influential politicians and members of the media, articles in national magazines, and a film titled *Wilderness River Trail.* As part of the compromise that saved the canyons at Dinosaur National Monument, the Sierra Club agreed not to challenge other dams in the project. One of those, the Glen Canyon Dam in Arizona, flooded a canyon that turned out to be a magnificent slickrock gorge that Brower and others believed, in retrospect, should have been saved as well. This was another hard lesson learned by the Sierra Club, and one that made club members and leaders suspicious of compromises.

Brower led the Sierra Club for sixteen years, through a period of major expansion. He opened the membership rolls and actively sought new members from the public. He developed a publications program, putting the Sierra Club name on a series of calendars and coffee-table books that combined beautiful photography with messages of preservation for specific areas. To prevent yet another dam, this time in the Grand Canyon, Brower used full-page advertisements in *The New York Times* to enlist the general public in a letter-writing campaign to dissuade Congress from authorizing the project.

The Sierra Club was also instrumental in creating the field of environmental law. For many years, conservation organizations had a difficult time pursuing court cases regarding land use because they were deemed to have no "standing"—that is, no financial stake—in the decisions being made. In a case in New York State in the late 1960's, attorneys representing local conservation organizations and the Sierra Club won the right to claim standing in land-use cases based on recreational, conservational, and aesthetic interests. In 1971 the Sierra Club created the Sierra Club Legal Defense Fund (renamed Earthjustice in 1997), a legally and financially distinct organization that would represent the club, other conservation organizations, and individuals in environmental litigation.

By the beginning of the twenty-first century, its policies driven by the interests of its more than one million members, the Sierra Club had taken on issues of water and air pollution, recycling, nuclear energy, population, and global warming. Through public education, grassroots letter-writing campaigns, lobby-

*Naturalist John Muir, first president of the Sierra Club.* (Library of Congress)

ing, and litigation, the Sierra Club continues to affect environmental opinion and policy in the United States and, increasingly, the world.

*Joseph W. Hinton*

FURTHER READING

Bevington, Douglas. *The Rebirth of Environmentalism: Grassroots Activism from the Spotted Owl to the Polar Bear.* Washington, D.C.: Island Press, 2009.

Cohen, Michael P. *The History of the Sierra Club, 1892-1970.* San Francisco: Sierra Club Books, 1988.

Jones, Holway R. *John Muir and the Sierra Club: The Battle for Yosemite.* San Francisco: Sierra Club Books, 1965.

McCloskey, J. Michael. *In the Thick of It: My Life in the Sierra Club.* Washington, D.C.: Island Press, 2005.

Miller, Norman. *Environmental Politics: Stakeholders, Interests, and Policymaking.* 2d ed. New York: Routledge, 2009.

Turner, Tom. *Sierra Club: One Hundred Years of Protecting Nature.* San Francisco: Sierra Club Books, 1991.


# Sierra Club v. Morton

CATEGORIES: Treaties, laws, and court cases; preservation and wilderness issues

THE CASE: U.S. Supreme Court decision regarding the principle of standing in U.S. environmental law

DATE: Decided on April 19, 1972

SIGNIFICANCE: *Sierra Club v. Morton* opened the federal courts to a wealth of environmental litigation because it settled the question about whether an injury to a noneconomic interest could provide the basis for a challenge to a federal agency decision.

During the early 1970's the Sierra Club sued the U.S. Forest Service to prevent the agency from approving permits that would allow Walt Disney Enterprises to construct a $35 million complex of motels, restaurants, swimming pools, and ski trails in Mineral King Valley, a quasi-wilderness area located in the Sierra Nevada of California. Up to fourteen thousand visitors per day were expected to gain access to the resort by using a 32-kilometer (20-mile) highway to be built, in part, through Sequoia National Park. The federal district court granted an injunction, but the federal court of appeals reversed it. The Supreme Court did not consider whether the building of the proposed development would violate federal law; rather, its focus was on whether the Sierra Club, as an organization with a special interest in the preservation of national parks and forests, had standing to challenge a federal agency's decision to issue the permits.

The Supreme Court had addressed the standing issue in *Association of Data Processing Service Organizations v. Camp* (1970), in which it had held that people who seek judicial review of a federal agency's action under Section 10 of the Administrative Procedure Act have to claim that the agency caused them injury in fact and that the injury was a harm within the zone of interests protected or regulated by statutes the agency was said to have violated. In *Sierra Club v. Morton*, Justice Potter Stewart's opinion for the Court addressed only the "injury in fact" element of the *Data Processing* test. The Court accepted the Sierra Club's claim that noneconomic injury constitutes injury in fact, that the "change in the aesthetics and ecology" caused by Mineral King's development "would destroy or otherwise adversely affect the scenery, natural and historic objects and wildlife of the park and would impair the enjoyment of the park for future generations." However, the Court rejected the Sierra Club's argument that it did not have to claim that its members would be adversely affected by the Mineral King development because the club's long-standing concern for and expertise in environmental matters gave it standing as a "representative of the public." In denying standing, the Court held that an organization's sincere interest in an environmental problem, even if the interest is of long duration and the organization is highly qualified to speak on behalf of the public, is not enough to satisfy the "injury in fact" requirement. If it were, the Court feared, there would be "no objective basis on which to disallow a suit by any other bona-fide organization no matter how small or short-lived."

Justice William O. Douglas, in an eloquent dissent, argued that the case should have been entitled *Mineral King v. Morton* and that the Court should have designed a standing rule that would have permitted the Sierra Club to litigate the Forest Service use permit on behalf of the valley. Drawing upon and citing Christopher Stone's law review article "Should Trees Have Standing?," Douglas proceeded to sketch the broad outlines of an imaginative redefinition of standing. The law, he said, indulges a fiction that inanimate objects such as ships and corporations are people and may, therefore, be parties to litigation. So

> ## Justice Douglas's Dissent
>
> *In his dissenting opinion in the case of* Sierra Club v. Morton, *Justice William O. Douglas argued for the legal standing of environmental objects:*
>
> The critical question of "standing" would be simplified and also put neatly in focus if we fashioned a federal rule that allowed environmental issues to be litigated before federal agencies or federal courts in the name of the inanimate object about to be despoiled, defaced, or invaded by roads and bulldozers and where injury is the subject of public outrage. Contemporary public concern for protecting nature's ecological equilibrium should lead to the conferral of standing upon environmental objects to sue for their own preservation.... This suit would therefore be more properly labeled as *Mineral King v. Morton.*
>
> Inanimate objects are sometimes parties in litigation. A ship has a legal personality, a fiction found useful for maritime purposes. The corporation sole—a creature of ecclesiastical law—is an acceptable adversary and large fortunes ride on its cases. The ordinary corporation is a "person" for purposes of the adjudicatory processes, whether it represents proprietary, spiritual, aesthetic, or charitable causes.
>
> So it should be as respects valleys, alpine meadows, rivers, lakes, estuaries, beaches, ridges, groves of trees, swampland, or even air that feels the destructive pressures of modern technology and modern life. The river, for example, is the living symbol of all the life it sustains or nourishes—fish, aquatic insects, water ouzels, otter, fisher, deer, elk, bear, and all other animals, including man, who are dependent on it or who enjoy it for its sight, its sound, or its life. The river as plaintiff speaks for the ecological unit of life that is part of it. Those people who have a meaningful relation to that body of water—whether it be a fisherman, a canoeist, a zoologist, or a logger—must be able to speak for the values which the river represents and which are threatened with destruction....
>
> The voice of the inanimate object, therefore, should not be stilled. That does not mean that the judiciary takes over the managerial functions from the federal agency. It merely means that before these priceless bits of Americana (such as a valley, an alpine meadow, a river, or a lake) are forever lost or are so transformed as to be reduced to the eventual rubble of our urban environment, the voice of the existing beneficiaries of these environmental wonders should be heard.
>
> Perhaps they will not win. Perhaps the bulldozers of "progress" will plow under all the aesthetic wonders of this beautiful land. That is not the present question. The sole question is, who has standing to be heard?

should it be with valleys, such as Mineral King, and with lakes, rivers, and forests. Who should speak for these inanimate objects and defend their rights? Congress, he argued, is "too remote . . . and too ponderous." Federal agencies, including the Forest Service, "are notoriously under the control of powerful interests." Only those who have an intimate relationship with valleys, lakes, rivers, and forests, because they hike, fish, or "merely sit in solitude and wonderment," may speak for the values that these natural objects represent.

Justice Harry Blackmun's dissent was much more direct in its criticism of the Court's "practical" decision. The Court's "somewhat modernized" conception of standing, he argued, was not adequate to deal with the novel issues raised by the deteriorating state of the environment. He suggested two alternatives: Either approve the district court's decision on the condition that the Sierra Club amend its complaint to comply with the Court's standing rule or redefine standing, as Justice Douglas had, to permit any bona fide environmental organization, such as the Sierra Club, to litigate on behalf of Mineral King. He did not fear, as the majority on the Court did, that an expanded definition of standing would open a Pandora's box of litigation, and he had much greater faith that appropriate restraints could be imposed on an "imaginative expansion" of standing.

*Sierra Club v. Morton* opened the federal courts to a wealth of environmental litigation because it settled the question left open by the *Data Processing* case about whether an injury to a noneconomic interest could provide the basis for a challenge to a federal agency decision. The Court further broadened its standing test in *United States v. Students Challenging Regulatory Agency Procedures* (1973) and *Duke Power v. Carolina Environmental Study Group* (1978) by allowing environmental groups to gain standing based on tenuous claims of causation between a proposed federal agency action and fairly speculative injuries. In *Lujan v. National Wildlife Federation* (1990), the Court tightened up standing and made it more difficult for environmental groups to gain access to federal courts and challenge federal programs by requiring them to allege that the specific lands involved were actually used by their members.

*William Crawford Green*

FURTHER READING

Buck, Susan J. *Understanding Environmental Administration and Law.* 3d ed. Washington, D.C.: Island Press, 2006.
Cox, Robert. "Public Participation in Environmental Decisions." In *Environmental Communication and the Public Sphere.* 2d ed. Thousand Oaks, Calif.: Sage, 2010.
Findley, Roger W., and Daniel A. Farber. *Environmental Law in a Nutshell.* 7th ed. St. Paul, Minn.: Thomson/West, 2008.
Hoban, Thomas, and Richard Brooks. *Green Justice: The Environment and the Courts.* 2d ed. Boulder, Colo.: Westview Press, 1996.
Stone, Christopher D. *Should Trees Have Standing? Law, Morality, and the Environment.* 3d ed. New York: Oxford University Press, 2010.

# Soil conservation

CATEGORIES: Agriculture and food; resources and resource management
DEFINITION: Use of agricultural and other cultivation practices aimed at maintaining soil quality and reducing erosion
SIGNIFICANCE: By using agricultural methods that protect soil from degradation, farmers can reduce erosion, prevent soil particles from contributing to air and water pollution, and improve crop production.

According to the United Nations Environment Programme, approximately 17 percent of the earth's vegetated land is degraded, a situation that poses a threat to agricultural production around the world. The introduction of minerals, metals, nutrients, fertilizers, pesticides, bacteria, and pathogens suspended in topsoil runoff into waterways is a significant source of water pollution and is a threat to fisheries, wildlife habitats, and drinking-water supplies. The introduction of soil particles into the air through wind erosion is a significant source of air pollution. Soil conservation is the effort by farmers and other land users to prevent the loss of topsoil from wind erosion, water erosion, desertification, and chemical deterioration such as the buildup of salts and fertilizer acids.

The Industrial Revolution of the nineteenth century and the population explosion of the twentieth century encouraged people to till new land, cut down forests, and disturb soil for the expansion of towns and cities. The newly exposed topsoil quickly succumbed to erosion from rainfall, floods, wind, ice, and snow. The Dust Bowl, which occurred in the Great Plains in the United States during the 1930's, is one example of the devastating effects of wind erosion.

Hugh Hammond Bennett, often called the father of soil conservation, lobbied for congressional establishment of the Soil Erosion Service, which was formed in the U.S. Department of the Interior in 1933, and the establishment of voluntary Soil Conservation Districts in each state. Bennett was named the first chief of the renamed Soil Conservation Service, now part of the Department of Agriculture, in 1937 (in 1994 the name of the agency changed again, to the Natural Resources Conservation Service). On August 4, 1937, the Brown Creek Conservation District in Bennett's home county, Anson County, North Carolina, became the first Soil Conservation District in the United States. Local landowners voted to establish the district by three hundred to one, proving that farmers were concerned about soil conservation. A reporter for the *Charlotte Observer* newspaper sought out the one negative voter, and after the program was explained to him, he changed his opinion. By 1948 more than 2,100 districts had been established nationwide; this number grew to 3,000 by the early years of the twenty-first century. The districts were eventually renamed Soil and Water Conservation Districts.

The Food Security Act of 1985 authorized the Conservation Reserve Program to take out of production any land deemed to be highly susceptible to erosion; it also required farmers to develop soil conservation plans for any remaining susceptible land. The Natural Resources Conservation Service has estimated that with such soil conservation measures, the loss of topsoil in the United States was cut nearly in half, reduced from 1.6 billion tons per year to 0.9 billion tons. The European Community and Australia also adopted soil conservation measures during the 1990's.

Soil conservation practices include covering the soil

with vegetation, reducing soil exposure on tilled land, creating wind and water barriers, and installing buffers. Vegetative cover slows the wind at ground level, slows water runoff, protects soil particles from being detached, and traps blowing or floating soil particles, chemicals, and nutrients. Because the greatest wind and water erosion damage often occurs during seasons in which no crops are growing or natural vegetation is dormant, soil conservation often depends on permitting the dead residues and standing stubble of the previous crop to remain in place until the next planting time. In forested areas, annual tree foliage loss serves as a natural ground mulch. Farmers can also reduce erosion by planting grass or legume cover crops until the next planting season for their primary crops or as part of a crop rotation cycle or no-till planting system.

Modern no-till and mulch-till planting systems reduce soil exposure to wind and rain. No-till systems leave the soil cover undisturbed before planting; crop seeds are inserted into the ground through narrow slots in the soil. In mulch-till planting, a high percentage of the dead residues of previous crops are retained on the soil's surface when a new crop is planted.

The ways in which crops are planted can also help to reduce erosion. Row crops can be planted at right angles to the prevailing winds and to the slope of the land in order to absorb wind and rainwater runoff energy and trap moving soil particles. Crops may be planted in small fields to prevent the avalanching caused by an increase in the amount of soil particles transported by wind or water as the distance across bare soil increases. As the amount of soil moved by wind or water increases, the erosive effects of the wind and water also increase. Smaller fields reduce the length and width of unprotected areas of soil.

Wind and water barriers include tree plantings and crosswind strips of perennial shrubs and tall grasses, which act as windbreaks, slowing wind speeds at the surface of the soil. The areas protected by such windbreaks extend for ten times the height of the barriers. In alley cropping, which is used in areas of sustained high winds, crops are planted between rows of larger mature trees. Contour strip farming on slopes, planting grass waterways in areas where rainwater runoff concentrates, and planting grass field borders 3 meters (10 feet) wide on all edges of cultivated or disturbed soil are additional methods for reducing wind speed and rainwater runoff and trapping soil particles, chemicals, and nutrients.

Soil conservation buffers work to filter agricultural runoff to remove sediments and chemicals. Riparian buffers are waterside plantings of trees, shrubs, and grasses, usually 6 meters (20 feet) in width. Riparian buffers planted only in grass are called filter strips. Grassed waterways, field borders, water containment ponds, and contour grass strips are other types of soil conservation buffers.

*Gordon Neal Diem*

FURTHER READING

Blanco, Humberto, and Rattan Lal. *Principles of Soil Conservation and Management.* New York: Springer, 2008.

Field, Harry L., and John B. Solie. "Erosion and Erosion Control." In *Introduction to Agricultural Engineering Technology: A Problem Solving Approach.* 3d ed. New York: Springer, 2007.

Plaster, Edward. *Soil Science and Management.* 5th ed. Clifton Park, N.Y.: Delmar Cengage Learning, 2008.

Schwab, Glen, et al. *Soil and Water Conservation Engineering.* 5th ed. Clifton Park, N.Y.: Delmar Cengage Learning, 2005.

## Sustainable agriculture

DEFINITION: The growing and harvesting of crops in a manner that has minimal impact on the environment

SIGNIFICANCE: The practices associated with sustainable agriculture help to protect the environment by preventing soil loss through erosion; minimizing the use of pesticides and chemical fertilizers, which can cause water pollution; conserving water; and enriching nutrient-depleted soils.

Most twentieth century agricultural practices were based on continued economic growth. This practice demonstrated dramatic increases in production but had negative impacts on the environment through the losses of plant and animal habitats, depletion of soil nutrients, and pollution of water supplies. The concept of sustainable development focuses on the use of renewable resources and working in harmony with existing ecological systems. The World Commission on Environment and Development described sustainable development as the ability "to

*Terraced rice fields in Guilin, China.* (©Shariff Che'Lah/Dreamstime.com)

meet the needs of the present without compromising the ability of future generations to meet their own needs." Practitioners of sustainable agriculture strive to manage their agricultural activities in such a way as to protect air, soil, and water quality, as well as conserve wildlife habitats and biodiversity.

Sustainable agriculture tries to match crops and livestock to the topography, soil characteristics, and climatic conditions that exist in a given region. The crops selected for cultivation must be well suited to the existing soil and site conditions and should also be resistant to known pests in the area.

### Problems Caused by Agriculture

Water pollution is one of the most damaging and widespread effects of modern agriculture. The runoff from farms accounts for more than 50 percent of sediment damage to natural waterways, and the chemicals and nutrients associated with this runoff in the United States are estimated to cost between $2 billion and $16 billion per year to clean. Heavy application of nitrogen fertilizers, insecticides, and herbicides has raised the potential for groundwater contamination. Feedlots that concentrate manure production lead to further groundwater contamination. Several of the most commonly used pesticides have been detected in the groundwater of at least one-half of U.S. states. In addition, monoculture, or the growing of a single crop on an area of land, requires heavy reliance on agricultural chemicals, because the practice of growing the same crop repeatedly depletes the natural organic nutrients that were formerly rich in North American topsoils.

Research has found that many of chemical agents, pesticides, fertilizers, plant-growth regulators, and antibiotics used in agriculture have ended up in the food supply. These chemicals can be harmful to humans at moderate doses, and chronic effects can develop with prolonged exposure at lower doses. Further, widespread pesticide use has been shown to result in severe stress in other animal populations, including bees. Pesticide use is often followed by occurrences of secondary pest outbreaks and resurgence of pests that have developed resistance to the pesticides previously used.

Because of these growing problems, many American farmers have turned to the practices associated with sustainable agriculture. The U.S. government offers guidance for this transition through the Food, Agriculture, Conservation, and Trade Act of 1990. This legislation describes sustainable agriculture as agriculture that, through an integrated system of plant and animal production practices, can, over the long run, meet human food and fiber needs, enhance environmental quality and natural resources, make the most efficient use of nonrenewable resources, maintain economic viability of farm operations, and enhance the quality of life for farmers and society.

### Water and Soil Conservation

In the western United States, the ability to irrigate crops is an especially important factor in agriculture, as much of the region is naturally arid or semiarid. In California, the limited supply of surface water has caused overdraft of groundwater and the consequent intrusion of salt water, which causes the permanent collapse of aquifers. In order to counteract such negative effects, sustainable farmers in California have introduced improved water conservation and storage methods, selecting drought-resistant crop species, using reduced-volume irrigation systems, and managing crops to reduce water loss. Drip and trickle irrigation methods also dramatically reduce water usage and water loss while helping to avoid such problems as soil salinization.

Farmers can temporarily manage salinization and contamination of groundwater with pesticides, nitrates, and selenium by using tile drainage to remove water and salt. This method often has adverse effects on the environment, however. Long-term solutions include conversion from the planting of row crops to production of drought-tolerant forages and the restoration of wildlife habitats.

One of the most important aspects of sustainable agriculture is soil conservation. In order to prevent excessive erosion, farmers might leave grass strips in the waterways of their fields to capture soil that begins to erode. A field with a 5 percent slope has three times the water runoff volume of a field with a 1 percent slope and eight times the soil erosion rate. Contour plowing, which involves plowing across a hill rather than up and down the hill, helps capture overland flow and reduce water runoff. Contour plowing is often combined with strip farming, where different kinds of crops are planted in alternating strips along the contours of the land. As one crop is harvested, another is still growing and helps recapture wind- and waterborne soil while preventing runoff from flowing quickly downhill. In areas of heavy rainfall, sustainable farmers might construct tiered ridges to trap water and prevent runoff. This method involves a series of ridges that are constructed at right angles to one

*Sunflowers and proso millet planted in alternating crop rotation plots. (USDA/David Nielsen)*

another to direct runoff and slow it down so that the water has a chance to soak into the soil.

Another farming method that contributes to soil conservation is terracing, in which the land is shaped into level shelves of earth to hold in water and soil. Soil-anchoring plants are grown on the edges of the terraces to provide further stability for the soil. Terracing is costly, but it can make enable farmers to grow crops on steep hillsides that they otherwise could not use for production. To protect fragile or unstable soil on sloping sites or along waterways, farmers may need to plant perennial species of grasses every year.

LIVESTOCK AND ANIMAL MANURE

Ruminant animals (sheep, cattle, and goats) can be raised on rangeland, pasture, cultivated forage, cover crops, shrubs, weeds, and crop residues. The breeds that have lower growth and milk production potential can adapt better to environments with sparse or seasonal forage. By growing row crops on level soil and growing pasture on steeper slopes, farmers can help reduce soil erosion; growing pasture and forage crops in rotation can help improve soil quality. When ruminants and other farm animals are allowed to graze in pasturelands, those fields are fertilized naturally. Farmers can direct how their fields are fertilized with animal manure by using portable fencing to keep livestock grazing in one area or strip of pasture all the way down before moving the animals to another strip of the field while the first strip of pasture recovers.

Sustainable farmers also use so-called green manure—crops that are raised specifically to be plowed under—to introduce organic matter and nutrients into the soil. Green manure crops help protect against erosion, cycle nutrients from lower levels of the soil into the upper layers, suppress weeds, and keep nutrients in the soil rather than allowing them to leach out. Legumes such as sweet clover, ladino clover, and alfalfa are excellent green manure crops. They are able to extract nitrogen from the air into the soil and leave a supply of nitrogen for the next crop that is grown. Some crops, such as beans and corn, can cause high soil erosion rates because they leave the ground bare most of the year. One way sustainable farmers combat this is by leaving crop residues on the land after harvest. Residues help reduce soil erosion and even excessive soil temperatures in hot climates. Many farmers choose to use cover crops rather than residue crops. The decision about which cover crop to grow is based on the farm's geographical location and the purpose of the crop: to control erosion, to capture nitrogen for the soil, to release nitrogen to the crop, or to improve soil structure and suppress weeds.

COVER CROPS

Cover crops such as hairy vetch or clover are well suited to the needs of later crops with high nitrogen requirements, such as tomatoes or sweet corn. Both of these cover crops decompose and release nutrients into the soil within one month. To fight erosion, a farmer might choose a fast-growing cover crop, such as rye. Rye provides abundant ground cover and an extensive root system that can prevent soil erosion and capture nutrients. Alfalfa, rye, or clover can be planted after harvest to protect the soil and add nutrients and can then be plowed under at planting time to provide a green manure for the crop. Cover crops can also be flattened with rollers, and seeds can be planted in their residue. This gives the new young plants a protective cover and discourages weeds from overtaking them. Use of natural nitrogen also reduces the risk of water contamination by agricultural chemicals.

Sustainable agriculture emphasizes the use of reduced tillage systems, which are intended to disturb the soil as little as possible when preparing it for planting. Minimum-till farming involves using the disc of a chisel plow to make a trench in the soil where seeds are planted. Plant debris is left on the surface of the ground between the rows, which helps prevent erosion. Conservation tillage, or conser-till farming, uses a cutting tool (a coulter) attached to a plow to open slots in the ground just wide enough to insert seeds without disturbing the soil. No-till planting involves drilling seeds into the ground directly through any ground cover or mulch.

CROP ROTATION AND MONOCULTURE

Planting the same crop every year on a given field can result in depleted soil. In order to keep the soil fertile, practitioners of sustainable agriculture rotate nitrogen-depleting crops (such as sweet corn, tomatoes, and cotton) from year to year with legumes, which add nitrogen to the soil. By planting winter cover crops, such as rye grass, farmers can protect their land from erosion. Such cover crops will, when plowed under, provide nutrient-rich soil for the planting of cash crops. Crop rotation also improves the physical condition of the soil because the different crops vary in root depth and in the ways they are cultivated.

In nature, various kinds of plants grow in mixed meadows, and this helps them to avoid insect infestations. The agricultural practice of monoculture places great quantities of the food of choice in easy proximity to insect predators. Insect populations that feed on particular crops can multiply more than they otherwise would when those crops are grown in the same fields year after year. Since most insects are instinctively drawn to the same home area every year, they cannot proliferate and thrive if their crops of choice are not in the same fields every year.

In addition to helping farmers use fewer pesticides, crop rotation helps to control weeds naturally. Some crops and cultivation methods inadvertently allow certain weeds to thrive. Sustainable farmers often incorporate into their crop rotations successor crops that eradicate weeds. Some crops, such as potatoes and winter squash, work as cleaning crops because of the different style of cultivation that is used on them. Pumpkins planted between rows of corn will help keep weeds at bay.

INTEGRATED PEST MANAGEMENT

Most sustainable farmers control insect pests through the practice known as integrated pest management (IPM). In IPM, each crop and its pests are evaluated as an ecological system. A plan is developed to manage the damage that pests can do through the use of particular cultivation techniques, biological methods, and chemical methods at different timed intervals. Although effective, profitable, and safe, IPM techniques have been adopted widely only for a few crops, such as tomatoes, citrus, and apples.

The goal of IPM is to keep pest populations below the size where they can cause damage to crops. Fields are monitored to gauge the level of pest damage. If farmers begin to see crop damage, they put cultivation and biological methods into effect to control the pests; physical techniques such as vacuuming bugs off crops are also used. IPM encourages the growth and diversity of beneficial organisms that enhance the defenses and vigor of plants. Small amounts of pesticides are used only if all other methods fail to control pests. It has been found that IPM, when done properly, can reduce inputs of fertilizer, lower the use of irrigation water, and reduce preharvest crop losses by 50 percent. Reduced pesticide use can cut the costs of pest control by 50 to 90 percent and can increase crop yields without increasing production costs.

*Toby Stewart and Dion Stewart*

FURTHER READING

Chiras, Daniel D. "Creating a Sustainable System of Agriculture to Feed the World's People." In *Environmental Science*. 8th ed. Sudbury, Mass.: Jones and Bartlett, 2010.

Francis, Charles A., Raymond P. Poincelot, and George W. Bird, eds. *Developing and Extending Sustainable Agriculture: A New Social Contract*. Binghamton, N.Y.: Haworth Press, 2006.

Gliessman, Stephen R., and Martha Rosemeyer, eds. *The Conversion to Sustainable Agriculture: Principles, Processes, and Practices*. Boca Raton, Fla.: CRC Press, 2010.

Koepf, Herbert H. *The Biodynamic Farm: Agriculture in the Service of the Earth and Humanity*. Hudson, N.Y.: Anthroposophic Press, 2006.

Lyson, Thomas A. *Civic Agriculture: Reconnecting Farm, Food, and Community*. Lebanon, N.H.: Tufts University Press, 2004.

Troeh, Frederick R., and Louis M. Thompson. *Soils and Soil Fertility*. 6th ed. Malden, Mass.: Blackwell, 2005.

# Sustainable development

CATEGORY: Resources and resource management

DEFINITION: Development that meets the consumption needs of the current generation without compromising the ability of future generations to increase their economic production to meet future needs

SIGNIFICANCE: When the principles of sustainable development are followed, environmental benefits arise as a consequence of changes in human attitudes and behaviors, resource utilization, and applications of technology.

According to the 1987 report of the United Nations World Commission on Environment and Development, also known as the Brundtland Commission, humanity has the ability to make development sustainable—to ensure that the current generation meets the needs of the present without compromising the ability of future generations to meet their own needs. Sustainable development involves a process of change in which the exploitation of resources, the direction of investments, the orientation of technological development, and institutional change are all in harmony and enhance both current and future potential to meet human needs and aspirations. The Brundtland Commission envisioned the possibility of continued economic growth, population stabilization, improvements in global economic equity among all nations, and environmental improvement, all occurring simultaneously and in harmony. Since publication of the commission's report, titled *Our Common Future*, the goal of sustainable development—both environmental and economic development—has become the dominant global position.

Advocates of sustainable development hold a normative philosophy, or value system, concerned with equal distribution of the earth's natural capital among current and future generations of humans. They promote three core values. First, current and future generations should have equal access to the planet's life-support systems—including the earth's gaseous atmosphere, biodiversity, stocks of exhaustible resources, and stocks of renewable resources—and should maintain the earth's atmosphere, land, and biodiversity for future generations. Exhaustible resources, such as minerals and fossil fuels, should be used sparingly and conserved for use by future generations. Renewable resources, such as forests and fertile soil, should be renewed as they are used to ensure that stocks are maintained at or above current levels and are never exhausted.

Second, all future generations should have the opportunity to enjoy a material standard of living equivalent to that of the current generation. In addition, the descendants of the current generation in underdeveloped regions should be permitted to increase their economic development to match that available to descendants of the current generation in the industrialized regions. Future development and growth in both developed and underdeveloped regions must be sustainable.

Finally, future development must no longer follow the growth path taken by the currently industrialized countries but should utilize appropriate technology. Development should also limit use of renewable resources to each resource's maximum sustained yield—that is, the rate of harvest of natural resources such as fisheries and timber that can be maintained indefinitely through active human management of those resources.

Weak sustainability requires that depletions in natural capital be compensated for by increases in human-made capital of equal value. For example, the requirements for weak sustainability are met when a tree (natural capital) is cut for the construction of a frame house (human-made capital). However, if the tree is cut and cast aside in a land-clearing project, the requirements for weak sustainability are not met. Strong sustainability requires that depletions of one sort of natural capital be compensated for by increases in the same or similar natural capital. For example, the requirements for strong sustainability are met when a tree is cut and a new tree is planted to replace it, or when loss of land in equatorial rain forests in Brazil is compensated for by an increase in the area of temperate rain forests on the Pacific coast of North America.

Sustainable development is promoted through a combination of public policies. First, to the extent possible, policy makers assign monetary values to elements in the earth's support system so that they can make the economic and financial calculations necessary to ensure that the requirements of weak sustainability are met. Second, economic development in the underdeveloped world is shifted away from high-resource-using, high-polluting patterns that have been seen in the developed nations and toward more sustainable or "appropriate" patterns. Suggested appropriate technologies and techniques for sustainability include solar energy, resource recycling, cottage industry, and microenterprises (factories built on a small scale). Third, objective and measurable quality standards for air, water, and other resources are established and enforced to ensure that a continuing minimum quality and quantity of natural capital is maintained and that certain stocks of natural capital are protected through the establishment of wilderness areas, oil and gas reserves, and other reserves. Finally, each individual human is encouraged to make a minimal personal impact on the earth's natural capital by adopting a commitment to a sustainable lifestyle.

Environmental improvement results from the changes in resource utilization that are part of sustainable development. For example, reductions in the use and waste of natural capital reduce the environmental impacts of resource extraction techniques such as strip mining and waste disposal methods such as incineration. The setting of environmental quality standards and policies requiring the maintenance of biodiversity lead to the implementation of antipollution efforts and ecosystem restoration projects.

*Gordon Neal Diem*

Further Reading

Bowers, John. *Sustainability and Environmental Economics: An Alternative Text.* Essex, England: Longman, 1997.

Dryzek, John S. "Environmentally Benign Growth: Sustainable Development." In *The Politics of the Earth: Environmental Discourses.* 2d ed. New York: Oxford University Press, 2005.

Landon, Megan. *Environment, Health, and Sustainable Development.* New York: Open University Press, 2006.

Lee, Kai N. *Compass and Gyroscope: Integrating Science and Politics for the Environment.* Washington, D.C.: Island Press, 1993.

Rogers, Peter P., Kazi F. Jalal, and John A. Boyd. *An Introduction to Sustainable Development.* Sterling, Va.: Earthscan, 2008.

Sitarz, Daniel, ed. *Agenda 21: The Earth Summit Strategy to Save Our Planet.* Boulder, Colo.: EarthPress, 1993.

## Sustainable forestry

CATEGORIES: Forests and plants; resources and resource management

DEFINITION: A system of management that relies on natural processes to maintain forests' continuing capacity to produce stable and perpetual yields of harvested timber and other benefits

SIGNIFICANCE: Sustainable forestry offers an environmentally sensitive alternative to the logging practice of clear-cutting and to the technique of monoculture tree farming, but disagreements exist among advocates of sustainable forestry regarding issues of ecosystem maintenance versus high timber yields.

Forest management in the United States first became an issue in 1827 when the Department of the Navy and President John Quincy Adams saw the need for a continuous supply of mature timber for ship construction. In the 1860's the American Association for the Advancement of Science first discussed the need for sustained-yield forestry. In 1878 the members of the Cosmos Club, a group of Washington, D.C., intellectuals, proposed the wise use of natural resources for the greatest good, for the greatest number, and for the longest time, establishing the foundation of the conservation movement. The first national forest reserves were established by the U.S. government in 1891, and the first selective logging and marketing of U.S. government timber reserves occurred in 1897. Clear-cutting was the general method of timber harvesting. Continued clear-cutting during the twentieth century resulted in deforestation of both private lands and lands overseen by the U.S. Forest Service, leading to concerns about soil erosion, water pollution, loss of wildlife habitat, and the sustained availability of forest resources.

Forestry science developed the system of high-yield plantation tree farming in the 1930's. By the 1960's ecological concerns had led to restoration forestry, which emphasizes human intervention to reconstruct forest ecosystems and return forests to baseline conditions that existed before clear-cutting or plantation planting. By the 1980's new understandings concerning the complexity of forest ecosystems led to an emphasis on perpetually sustaining existing forest resources rather than relying on human efforts to reconstruct forests.

Sustainable forestry is an alternative to clear-cutting, the standard logging practice. Clear-cutting removes all timber in one harvest; a given area is harvested usually no more than once every sixty to one hundred years. During clear-cutting, both mature and immature trees are removed in one process. Logging roads are cut into the forest so heavy machinery can remove all trees from a large area, usually about 40.5 hectares (100 acres) at a time. Clear-cutting and the accompanying road construction lead to soil erosion and nutrient loss, topsoil loss, silting and pollution of waterways, loss of wildlife habitat, and loss of recreational benefits. Repeated cycles of growth and clear-cutting erode soil nutrition; destroy plants, animals, and microorganisms in the ecosystem necessary for healthy forest growth; and reduce the value of future harvests.

Sustainable forestry is also an alternative to monoculture plantation forestry. Plantation forestry requires active human intervention to plant tree seedlings, control diseases and pests, and nurture timber stands to maturity. Plantations usually feature grid plantings of single tree species, with all trees maturing simultaneously. The lack of species and age diversity makes tree plantations unsuitable for wildlife habitat or recreation and also makes the trees susceptible to diseases and pests. Monoculture plantations also deplete species-specific minerals and other nutrients in the soil, reducing its future productivity.

Sustainable forest management techniques seek a

perpetual high yield of timber and pulpwood while maintaining biological diversity and natural forest ecosystems and permitting forests to restore their vitality through natural processes, such as foliage decomposition and fire. These techniques are designed to maintain a balance between natural environmental stresses and the human needs for timber, pulpwood, and a variety of harvested forest products, as well as recreation in natural settings. In spite of efforts to maintain this balance, however, various sustainable forestry methods often tend to favor either ecosystem maintenance or high timber yields.

Sustainable forestry with an ecosystem emphasis is the discipline of repeated thinning of natural tree stands to sustain a mixed-age, mixed-species forest that is naturally perpetuated by seeds from the mature trees. The forest is periodically thinned, usually every twenty years, to provide a steady income to the forest owners, permit the remaining trees to reach their full maturity, and provide space for new seedlings to grow. When the timber stand reaches full sustainable maturity, immature trees are continuously harvested for pulpwood, and mature trees more than one hundred years of age are continuously harvested for high-quality lumber. Natural processes promote the health of the forest and revitalize the forest soil. Diversity in both ages and species of trees makes the forest a suitable habitat for a variety of forest-dwelling animal species and for human recreation. The forest is able to recover quickly from natural disasters, fires, or drought.

Sustainable forestry with an emphasis on timber yield divides the forest into subplots, then manages each subplot to produce two sequential high-yield plantation crop cycles of eighty years each before permitting the plot to grow to maturity in a third four-hundred-year cycle. The third cycle permits the forest soil to restore its vitality and produces an old-growth forest suitable for wildlife and eventual timber harvesting. Once fully implemented, this system ensures that the forest has subplots at each stage of growth and harvesting, from newly planted plots to old-growth plots with trees at or near four hundred years of age.

*Gordon Neal Diem*

FURTHER READING

Berger, John J. *Forests Forever: Their Ecology, Restoration, and Protection.* Chicago: University of Chicago Press, 2008.

Bettinger, Pete, et al. *Forest Management and Planning.* Burlington, Mass.: Academic Press, 2009.

Colfer, Carol J. Pierce. *The Equitable Forest: Diversity, Community, and Resource Management.* Washington, D.C.: Resources for the Future, 2005.

Davis, Lawrence S., et al. *Forest Management: To Sustain Ecological, Economic, and Social Values.* 4th ed. Boston: McGraw-Hill, 2001.

List, Peter C., ed. *Environmental Ethics and Forestry: A Reader.* Philadelphia: Temple University Press, 2000.

Maser, Chris. *Our Forest Legacy: Today's Decisions, Tomorrow's Consequences.* Washington, D.C.: Maisonneuve Press, 2005.

## Tellico Dam

CATEGORY: Preservation and wilderness issues
IDENTIFICATION: Hydroelectric dam on the Little Tennessee River near Knoxville, Tennessee
DATE: Completed on November 29, 1979
SIGNIFICANCE: Controversy over the transformation of a river valley into an artificial lake brought national attention to the conflict between wilderness preservation and human development.

As early as 1936, the Tennessee Valley Authority (TVA) had made plans to build a dam across the Little Tennessee River to facilitate navigation below Fort Loudoun, but the project was vetoed in 1942 because of a scarcity of steel. Building the dam remained a high priority for the TVA, and in 1959 the agency undertook a thorough study of how the region would be affected by the dam. The study also compared the projected cost of the dam to the benefits it would create, which included electricity as well as employment opportunities and recreational sites. The cost was estimated to be about equal to the possible benefits, and the TVA decided that building the dam would be economically feasible and beneficial to the area.

The appropriation of federal money for the dam's construction had to be approved by the president of the United States. On October 17, 1966, the TVA received $3.2 million to begin building the dam in 1967, with the completion date estimated to be 1970 or 1971. The final cost of the dam was actually $120 million, and it was not completed until November 29, 1979. The delay and extra costs were caused by many factors, including inflation and the diversion of federal money to support the Vietnam War.

Aside from economic factors, however, the construction of the dam was also delayed during the 1970's by legal action taken against the project by the Cherokee Nation, local residents, and environmentalists. The Little Tennessee Valley, which would be flooded upon completion of the project, contained many Cherokee historical sites, including sacred burial grounds and ruins of the Seven Towns, which were the center of the Cherokee Nation before the Cherokees were sent to reservations in Oklahoma and North Carolina. Residents of the area that would be flooded asserted their interest in maintaining homes that had been in their families for generations.

Prime farmland would also be lost once the valley flooded. Environmentalists pointed out that the dam was unnecessary given that, in comparison with the TVA's total electrical output, the dam would produce very little electricity. Within 96 kilometers (60 miles) of Tellico, twenty-four major dams already existed. The environmentalists also argued that the recreational opportunities offered by a new lake, such as boating, swimming, and fishing, were trivial compared to the greater wilderness activities associated with an untamed river near Great Smoky Mountains National Park. Building the dam would restrict the last free-flowing stretch of the Little Tennessee River.

After the discovery of an endangered species of fish, the snail darter, in the Little Tennessee River in 1973, environmentalists filed suit in 1977 to halt construction of the dam because operation of the dam would destroy the fish's habitat. After many legal battles, however, a special law was passed to exempt the TVA from complying with the Endangered Species Act of 1973, and Tellico Dam went into operation in January of 1980.

*Rose Secrest*

### Further Reading

Murchison, Kenneth M. *The Snail Darter Case: TVA Versus the Endangered Species Act.* Lawrence: University Press of Kansas, 2007.

Palmer, Tim. "The Movement to Save Rivers." In *Endangered Rivers and the Conservation Movement.* 2d ed. Lanham, Md.: Rowman & Littlefield, 2004.

Wheeler, William Bruce, and Michael J. McDonald. *TVA and the Tellico Dam, 1936-1979: A Bureaucratic Crisis in Post-industrial America.* Knoxville: University of Tennessee Press, 1986.

# United Nations Convention to Combat Desertification

CATEGORIES: Treaties, laws, and court cases; resources and resource management

THE CONVENTION: International agreement intended to address the problems of land degradation and desertification around the world

DATE: Adopted on June 17, 1994

SIGNIFICANCE: The United Nations Convention to Combat Desertification has brought to the attention of decision makers some of the most vulnerable ecosystems in the world: drylands and deserts. The degradation of these ecosystems has affected some of the most vulnerable people on earth, causing massive migrations and creating environmental refugees. The convention has helped bring increased attention also to particular issues related to desertification, such as climate change and trade liberalization.

It has been widely acknowledged that land degradation and desertification are serious problems in many regions of the world, with significant impacts on the economic, social, cultural, and environmental well-being of the populations they affect. Scientists and policy makers had been talking about how to address the problems of land degradation and desertification for many years before the United Nations Conference on Desertification in 1977 finally began strong international efforts to address these problems.

Participants at the 1992 United Nations Conference on Environment and Development, known as the Earth Summit, in Rio de Janeiro, Brazil, discussed the issue of desertification and decided to call on the U.N. General Assembly to establish a committee to prepare a convention addressing desertification by June, 1994. On June 17, 1994, the United Nations Convention to Combat Desertification (UNCCD) was adopted in Paris and subsequently opened for signature on October 14 of the same year. The convention entered into force on December 26, 1996, after it was ratified by 50 countries. The first Conference of the Parties (COP), the convention's governing body, was held in Rome, Italy, in October, 1997. By 2009 the number of countries that were parties to the convention had grown to 193.

A permanent secretariat is responsible for providing support to affected UNCCD member countries, assisting in the preparation for sessions of the COP, distributing information as it becomes available, and coordinating programs and activities with relevant international nonprofit environmental organizations. The United Nations Global Environment Facility is the official financing mechanism of the UNCCD. Since 2001 the Committee for the Review of the Implementation of the Convention has assisted the COP in conducting regular reviews of signatory nations' implementation of the convention.

At the country level, the desertification issues that a member nation faces must be detailed in a National Action Program, along with the measures being taken to address them. The UNCCD's approach is to support sustainable development at the community level, with the reasoning that in turn this will lead to overall reductions in land degradation and the protection of fragile ecosystems.

The UNCCD recognizes that human activities are the primary causes of land degradation and desertification. It further recognizes that land degradation and desertification result from a number of complex factors—political, physical, economic, social, cultural, and biological—and the interactions among them. Factors such as poverty, lack of food security, and lack of proper nutrition also play important roles, influencing human activities in ways that often contribute to land degradation and desertification.

*Lakhdar Boukerrou*

FURTHER READING

Delville, Philippe L. *Societies and Nature in the Sahel.* London: Taylor & Francis, 2007.

Geist, Helmut. *The Causes and Progression of Desertification.* Burlington, Vt.: Ashgate, 2005.

Goudie, Andrew. "The Human Impact on Vegetation." In *The Human Impact on the Natural Environment: Past, Present, and Future.* 6th ed. Malden, Mass.: Blackwell, 2005.

Middleton, N. *Global Desertification: Do Humans Cause Deserts?* Amsterdam: Elsevier, 2004.

United Nations. Convention to Combat Desertification Secretariat. *Desertification: Coping with Today's Global Challenges.* Eschborn, Germany: Author, 2008.

Williams, M. A. J., and Robert C. Balling, Jr. *Interactions of Desertification and Climate.* London: Arnold, 1996.

## Watt, James

CATEGORY: Preservation and wilderness issues
IDENTIFICATION: American attorney who served as U.S. secretary of the interior
BORN: January 31, 1938; Lusk, Wyoming
SIGNIFICANCE: Watt was labeled a major antienvironmentalist during his tenure as secretary of the interior; he was frequently accused of using his office to weaken environmental policies that fell under his domain of authority.

James Watt's ancestors were nineteenth century homesteaders in Wyoming, laying claim to a large tract of land for ranching. Watt was both a rancher and a successful lawyer. As a child, he became familiar with the harsh, barren land of the family ranch, pumping water for the cattle, repairing fences, and performing other difficult chores. He recalled later in life that his early experiences trained him to challenge a hostile environment.

In 1962 Watt obtained a degree in law from the University of Wyoming; shortly thereafter, he moved to Washington, D.C., to assume a position as legislative assistant and counsel to Senator Milward Simpson of Wyoming. From 1966 to 1969 Watt worked with the U.S. Chamber of Commerce's Washington, D.C., office in natural resource and environmental pollution policy. He lobbied for prodevelopment business interests in such areas as the use of public lands for mining, energy, and water resource development. During the presidential administrations of Richard Nixon and Gerald Ford, Watt served in the Department of the Interior as an assistant secretary responsible for water and energy resources and as director of the Bureau of Outdoor Recreation.

During the 1970's Watt became closely associated with the Sagebrush Rebellion, a movement of ranchers and entrepreneurs in the American West who opposed numerous federal regulations that they asserted were inhibiting the profitable exploitation of natural resources. In 1977 Watt assumed the presidency of the Mountain States Legal Foundation, an organization founded by brewing magnate Joseph Coors to provide assistance for individuals who challenge government restrictions on strip mining, oil and gas exploration, mineral extraction, and grazing lands.

Watt was a logical choice to become secretary of

the interior in President Ronald Reagan's administration, which was committed to economic expansion and resource development with minimal government intrusion. From the beginning of his tenure in January, 1981, Watt worked to cut the department's budget and eliminate agencies and programs. He promoted measures to ease restrictions on oil and gas exploration on federal lands and in offshore waters, open more federal land for grazing and timber cutting, facilitate the construction of dams and reservoirs to improve irrigation of farmland, and restrict expansion of the national park system. Watt's actions were consistent with his ideology of economic growth with minimal government interference, and he felt compelled by his religious convictions to "follow the Scriptures which call upon us to occupy the land until Jesus returns."

Environmental groups strongly opposed Watt's appointment as secretary of the interior, arguing that his ideology was contrary to the mission of the department—to manage federal lands in the public interest. During his tenure as secretary, Watt refused to meet with environmental group leaders and made statements suggesting that they were subversive and were weakening the United States. He resigned from his position in November, 1983, after he was criticized for remarks he made about Senate members who had been appointed to a coal advisory panel. At the time of Watt's resignation, the Senate was working on a resolution calling for his dismissal.

*Ruth Bamberger*

### Further Reading

Andrews, Richard N. L. *Managing the Environment, Managing Ourselves: A History of American Environmental Policy.* 2d ed. New Haven, Conn.: Yale University Press, 2006.

Davis, Charles, ed. *Western Public Lands and Environmental Politics.* 2d ed. Boulder, Colo.: Westview Press, 2001.

Short, C. Brant. "Conservation Reconsidered: Environmental Politics, Rhetoric, and the Reagan Revolution." In *Green Talk in the White House: The Rhetorical Presidency Encounters Ecology,* edited by Tarla Rai Peterson. College Station: Texas A&M University Press, 2004.

# Wetlands

CATEGORY: Preservation and wilderness issues
DEFINITION: Transitional areas between terrestrial and aquatic ecosystems that exhibit characteristics of both
SIGNIFICANCE: Wetlands are widely considered to be among the world's most important ecosystems because of their high biodiversity and productivity. They also perform important functions related to the maintenance of surface and groundwater quality and quantity, prevention of saltwater intrusion, control of coastal erosion, and regulation of climate.

Wetlands are often distinguished by having three major components: water, hydrophytic vegetation, and hydric soils. All wetlands have water present for at least part of the year, though the depth and duration of flooding vary considerably. Some wetlands have water-saturated soil, whereas others are characterized by permanent flooding. At least periodically, wetlands support a predominance of hydrophytic vegetation— that is, plant life adapted to thrive in saturated soil conditions. Wetlands are also characterized by having undrained, or hydric, soils. These are soils in which an anaerobic condition (an absence of free oxygen) has developed because of long periods of saturation, flooding, or ponding during the growing season.

### Defining Wetlands

No single, formal definition of wetlands has been established, because no single description is appropriate for all of the diverse wetland types that exist over a large geographic scale with diverse climatic conditions. Dozens of definitions have been written, however, for specific reasons by specific interest groups and various regulatory agencies. The problem of defining wetlands is one of critical consequence to those persons who are subject to restrictions and limitations placed on them by various national, regional, state, and local laws concerning wetlands. Inconsistent definitions place a severe burden on private landowners who may be subject to such laws and do not have adequate technical or legal knowledge about wetlands.

In the United States, a regulatory definition of the term "wetland" has been developed so that the Army Corps of Engineers and the Environmental Protection Agency (EPA) can administer the permitting of dredging and filling of wetlands as prescribed in section 404 of the Clean Water Act. For this purpose, wetlands are considered to be

> those areas that are inundated or saturated by surface or ground water at a frequency and duration sufficient to support, and that under normal circumstances do support, a prevalence of vegetation typically adapted for life in saturated soil conditions. Wetlands generally include swamps, marshes, bogs, and similar areas.

Thus jurisdictional wetlands in the United States—those that are subject to section 404 permitting—must possess all three key characteristics: hydrology, hydrophytic vegetation, and hydric soils. However, because many wetlands are not permanently wet and because water may not be seen during a single site visit, positive hydrology indicators must be found, which must be supported by wetland vegetation and soils. These strict requirements for wetland identification have caused many wetlands to fall into uncertain categories. For example, some wetlands can have the appropriate hydrology but fail to develop the appropriate wetland soils and vegetation; often two of the three characteristics can be confirmed but not the third. These situations continue to cause confusion among governmental agencies and private landowners.

Another significant issue concerning wetlands is that they form ecotones (ecological transition zones) between upland and aquatic ecosystems. Thus even if an area is identified as a wetland, determination of its exact boundaries may be extremely difficult because of the gradual, perhaps imperceptible, changes in soil and vegetation characteristics. The problem of identifying areas as wetlands and defining the boundaries of those wetlands is known as wetland delineation. The ability to perform delineations is acquired only through extensive training, especially in the areas of soils and botany.

### Functions and Value

Not all wetlands perform the same set of functions or perform their functions at the same rate or efficiency. Often the size of a wetland and its location in the watershed determine its functions. Wetlands generally have extremely high biodiversity and rates of productivity. They provide food, shelter, and water for various invertebrates and vertebrates, many of which may be endangered or threatened. According to the U.S. Department of Agriculture's Natural Resources Conservation Service, nearly 5,000 plant species, 190 amphibian species, one-third of all bird species, and all wild ducks and geese depend on the nation's wetlands. Endangered species in the United States that have wetland habitats include the bald eagle, red wolf, whooping crane, fatmucket mussel, and swamp rose. It has been estimated that wetlands provide essential habitat for 40 percent of the nation's endangered and 60 percent of its threatened species.

Another function associated with wetlands is the maintenance of the quantity and quality of both surface water and groundwater. Many wetlands serve to recharge aquifers. Wetlands also accumulate sediments, nutrients, and many forms of water pollutants from their watersheds. By removing these materials, wetlands serve to clean the water.

Wetlands are sometimes referred to as nature's sponges because of their ability to ameliorate the effects of stormwater runoff and reduce floodwater damage. Stormwater enters wetlands and spreads out over large areas and then is slowly released. Increased property damage from flooding has been shown to occur following the destruction of wetlands. The Army Corps of Engineers noted in a 1976 report that if the Charles River wetlands near Boston, Massachusetts, were destroyed, flood damage in the river basin would increase by as much as $17 million annually. Other wetland functions are prevention of saltwater intrusion into groundwater and surface-water supplies, protection against coastal erosion from storms, and regional and global climate stabilization. Concerns regarding global climate change have sparked interest in the ability of wetlands to function as carbon reservoirs.

A wetland value is any product, characteristic, or function of a wetland that has worth or is beneficial to the environment or to people. Wetland products such as timber, fiber, food, and fish have commercial value and are easily measured. Some other values of wetlands, however, are not as easily quantified. For example, wetlands may have sociocultural significance and provide sites for recreation, research, and education. Further, it is impossible to assign a dollar value to the fact that wetlands provide habitat for a high percentage of endangered and threatened species around the world.

The value placed on a wetland's functions is, in many cases, the most important factor that determines whether the wetland is preserved or converted to some other use. As society's needs and perceptions change over time, the value assigned to wetland functions also changes. The values associated with wetlands are often in conflict because of the large number of functions these ecosystems can perform. For example, if the water level in a wetland is raised, waterfowl production may increase while timber production decreases. Managers of wetlands may thus make decisions that are popular with one user group but unpopular with other user groups.

### Loss and Degradation

"Wetland loss" refers to a decrease in wetland area caused by the conversion of wetland to nonwetland. "Wetland degradation" refers to the impairment of one or more wetland functions because of human activity. In most cases, wetland loss is difficult or impossible to reverse because of the complexity of wetland structure and function. Wetland degradation, in contrast, is more easily reversed through a variety of applied science and conservation tools. Wetland creation—the formation of wetlands in formerly nonwetland areas—has become an increasingly common strategy for combating wetland loss.

Wetlands are found on every continent. Even beneath Antarctica's mantle of ice, there are wetlands that support life. The exact size of global wetland areas is difficult to assess because of differences in wetland definitions and lack of documentation in many countries, but wetlands have been estimated to cover about 6 percent of the land area of the earth. The largest wetland areas are found in tropical, subtropical, and boreal regions. Since the beginning of the twentieth century, the world has lost more than 50 percent of its wetland area.

A 1990 report issued by the U.S. Fish and Wildlife Service estimated that the area of wetlands in the United States decreased from about 158 million hectares (391 million acres) during the 1780's to about 111 million hectares (274 million acres) in the mid-1980's—a 30 percent overall loss of wetland area. Within the lower forty-eight states, the estimated loss was 53 percent. The Fish and Wildlife Service has been monitoring the nation's wetlands trends since the 1950's. During the period from the mid-1950's to the mid-1970's an estimated 185,400 hectares (458,000 acres) of wetlands in the coterminous United States were lost every year. From the mid-1970's to the mid-1980's, by which time the value of wetlands had begun to be recognized, the estimated rate of loss decreased to 117,400 hectares (290,000 acres) per year. From 1986 to 1997 the loss rate dropped by 80 percent, to 23,700 hectares (58,500 acres) annually. Between 1998 and 2004 net wetlands gains in the coterminous United States surpassed net losses for the first time since the survey began, thanks to wetlands creation and restoration efforts. The net gain for the period was 77,598 wetland hectares (191,750 acres), or an average net gain of 12,950 hectares (32,000 acres) per year. As of 2004 there were an estimated 43.6 million hectares (107.7 million acres) of wetlands in the coterminous United States.

The underlying causes of wetland loss and degradation are numerous. These include poverty and economic inequality; population pressures from growth, immigration, and mass tourism; social and political conflicts; high demand for wetland resources such as timber; drainage; diking and damming; air and water pollution; introduction of exotic species; natural events such as hurricanes; and economic policies. In the United States, approximately 80 percent of wetland losses from the mid-1950's through the mid-1970's were the result of agricultural practices. Since the 1970's growing awareness of the importance of wetland functions has slowed the destruction of wetlands, but wetland loss and, more frequently, degradation continue, caused by development and other stressors.

### Conservation and Protection

Wetlands in the United States are protected through regulation, economic programs, and acquisitions. At the federal level, a confusing mix of programs and legislation simultaneously encourages and discourages wetland conservation. As early as 1903, President Theodore Roosevelt recognized that wetland loss had become significant. By executive order, he established Pelican Island in Florida as the nation's first wildlife refuge. The federal government also protects wetlands through several laws, including the Clean Water Act of 1972, which created a plan to control the discharge of dredged or fill materials into wetlands and other waters of the United States. The Army Corps of Engineers and EPA share responsibility for implementing the program. The "swampbuster" program is part of the Food Security Act of 1985 and 1990. It seeks to remove federal incentives for the agricultural conversion of wetlands to nonwetlands. In conjunction with this act, the 1990 Farm Bill created a voluntary Wetland Reserve Program, which provides financial incentives to farmers to restore and protect wetlands through the use of long-term easements.

The North American Waterfowl Management Program represents another milestone in the conservation of important wetland habitat. This plan was signed between Canada and the United States in 1986 to restore declining waterfowl populations through habitat acquisition, development of economic incentives to change land-use practices, and improvement of water management. Mexico became a signatory to the agreement in 1994. At the global level, the most significant wetland conservation work has resulted from the Convention on Wetlands of International Importance, or Ramsar Convention, in 1971. This global treaty provides a framework for the international protection and wise use of wetlands.

U.S. presidents became active in wetland protection during the 1970's. President Jimmy Carter signed two executive orders that provide guidance for wetland and floodplain management and protection of these areas by federal agencies. President George H. W. Bush extended these efforts to recommend that the United States establish a national goal of "no net loss" of wetlands. This policy became a major force for wetland conservation in the United States.

Despite all of this activity, several difficulties remain. Wetlands, because of the complexity of their values and functions, continue to be managed for a variety of purposes, many of them conflicting. The policy of no net loss of wetlands applies to the loss of wetland acreage only, not to wetland functions, values, or quality. President Bill Clinton took a compromise position in wetlands protection by reaffirming the "no net loss" policy and supporting the Wetland Reserve Program; however, he created section 404 exemptions for 21.4 million hectares (53 million acres) of previously converted wetlands and for small plots of land owned by families who wanted to build single-family houses.

On Earth Day 2004, President George W. Bush announced an initiative to achieve an overall increase in the quantity and quality of U.S. wetlands. While there was a net gain in wetland area between 1998 and 2004, there were still wetland losses; urban and rural development was responsible for about 61 percent of the net freshwater wetlands losses during this period. Two notable U.S. Supreme Court cases during the Bush administration, *Solid Waste Agency of Northern Cook County v. Army Corps of Engineers* (2001) and *Rapanos v. United States* (2006, consolidated with *Carabell v. Corps of Engineers*), found that the Clean Water Act did not protect the isolated wetlands addressed in each case from being developed.

*Roy Darville*
*Updated by Karen N. Kähler*

FURTHER READING

Batzer, Darold P., and Rebecca R. Sharitz, eds. *Ecology of Freshwater and Estuarine Wetlands.* Berkeley: University of California Press, 2006.

Fowler, Theda Braddock, and Lisa Berntsen. *Wetlands: An Introduction to Ecology, the Law, and Permitting.* 2d ed. Lanham, Md.: Government Institutes, 2007.

Keddy, Paul A. *Wetland Ecology: Principles and Conservation.* 2d ed. New York: Cambridge University Press, 2010.

Mitsch, William J., and James G. Gosselink. *Wetlands.* 4th ed. Hoboken, N.J.: John Wiley & Sons, 2007.

Spray, Sharon L., and Karen L. McGlothlin, eds. *Wetlands.* Lanham, Md.: Rowman & Littlefield, 2004.

Tiner, Ralph W. *Wetland Indicators: A Guide to Wetland Identification, Delineation, Classification, and Mapping.* Boca Raton, Fla.: CRC Press, 1999.

## Wild and Scenic Rivers Act

CATEGORIES: Treaties, laws, and court cases; preservation and wilderness issues
THE LAW: U.S. federal law establishing a system of free-flowing rivers for the protection of their natural, scenic, and recreational values
DATE: Enacted on October 2, 1968
SIGNIFICANCE: Since the late 1960's the Wild and Scenic Rivers Act has been an important legislative tool for protecting America's pristine and free-flowing rivers from impoundments and unrestricted development.

During the 1950's and 1960's federal agencies such as the Bureau of Reclamation and U.S. Army Corps of Engineers were engaged in hundreds of projects for building large and small river impoundments for irrigation, flood control, recreation, and hydroelectric power generation. Seeing the loss of free-flowing rivers, the Outdoor Recreation Resources Review Commission, a federal panel of resource experts,

## Wild and Scenic Rivers Act

*Congress passed the Wild and Scenic Rivers Act of 1968 to recognize and protect the rivers of the United States. Main provisions of the act follow.*

*Congressional Declaration of Policy.* It is hereby declared to be the policy of the United States that certain selected rivers of the Nation which, with their immediate environments, possess outstandingly remarkable scenic, recreational, geologic, fish and wildlife, historic, cultural, or other similar values, shall be preserved in free-flowing condition, and that they and their immediate environments shall be protected for the benefit and enjoyment of present and future generations. The Congress declares that the established national policy of dam and other construction at appropriate sections of the rivers of the United States needs to be complemented by a policy that would preserve other selected rivers or sections thereof in their free-flowing condition to protect the water quality of such rivers and to fulfill other vital national conservation purposes.

*Classification.* A wild, scenic or recreational river area eligible to be included in the system is a free-flowing stream and the related adjacent land area.... Every wild, scenic or recreational river in its free-flowing condition, or upon restoration to this condition, shall be considered eligible for inclusion in the national wild and scenic rivers system and, if included, shall be classified, designated, and administered as one of the following:

(1) *Wild river areas*—Those rivers or sections of rivers that are free of impoundments and generally inaccessible except by trail, with watersheds or shorelines essentially primitive and waters unpolluted. These represent vestiges of primitive America.

(2) *Scenic river areas*—Those rivers or sections of rivers that are free of impoundments, with shorelines or watersheds still largely primitive and shorelines largely undeveloped, but accessible in places by roads.

(3) *Recreational river areas*—Those rivers or sections of rivers that are readily accessible by road or railroad, that may have some development along their shorelines, and that may have undergone some impoundment or diversion in the past.

suggested that legislation was needed to create a river preservation system. Under the sponsorship of Senator Frank Church of Idaho, the Wild and Scenic Rivers Act was signed into law by President Lyndon B. Johnson in October, 1968. Segments of eight rivers received protection as part of the original legislation: the Middle Fork of the Clearwater River and the Middle Fork of the Salmon River in Idaho, the Eleven Point in Missouri, the Feather in California, the Rio Grande in New Mexico, the Rogue in Oregon, the St. Croix in Minnesota and Wisconsin, and the Wolf in Wisconsin.

A key provision of the act is the way new rivers can be added to the National Wild and Scenic Rivers System. To qualify for protection a river has to be free-flowing and must possess outstanding scenic, recreational, geological, or other qualities. In lieu of an entire river being designated, the act enables river segments, including tributaries, to be added through congressional action, by nomination by the governor of the state in which the river is located, or through recommendation of the secretary of the interior or the secretary of agriculture. Along with state governments, four federal agencies are responsible for managing rivers in the system: the National Park Service, the Bureau of Land Management, the Fish and Wildlife Service, and the Forest Service.

River segments are designated within one of three categories—wild, scenic, or recreational—corresponding to their level of development. Segments categorized as wild must be inaccessible except by trail and free of impoundments. Scenic segments must also be free of impoundments and must have shorelines that are mostly undeveloped, with accessibility to roads located in a minimum of locations. Segments designated as recreational are the most developed and may have experienced some type of impoundment in the past. The act also includes stipulations concerning how rivers should be managed. For example, river management should not interfere with public use or enjoyment and should be respectful of private property rights.

By 2010 more than 19,300 kilometers (12,000 miles) of 252 rivers located in thirty-eight states had become part of the National Wild and Scenic River System. This included approximately 9,800 kilometers (6,100 miles) of river classified as wild, 4,300 kilometers (2,700 miles) classified as scenic, and 5,800 kilometers (3,600 miles) designated recreational. In 1995 the Interagency Wild and Scenic River Coordinating Council was created to provide oversight for the system. The council is made up of representatives from all the federal agencies involved in managing the system's rivers.

*Thomas A. Wikle*

## Further Reading

Cech, Thomas V. "Water, Fish, and Wildlife." In *Principles of Water Resources: History, Development, Management, and Policy*. 3d ed. New York: John Wiley & Sons, 2010.

Echeverria, John D., Pope Barrow, and Richard Roos-Collins. *Rivers at Risk: The Concerned Citizen's Guide to Hydropower.* Washington, D.C.: Island Press, 1989.

Palmer, Tim. *Endangered Rivers and the Conservation Movement*. 2d ed. Lanham, Md.: Rowman & Littlefield, 2004.

River Network and David M. Bolling. *How to Save a River: A Handbook for Citizen Action*. Washington, D.C.: Island Press, 1994.

# Wilderness Act

CATEGORIES: Treaties, laws, and court cases; preservation and wilderness issues
THE LAW: U.S. federal legislation concerning the preservation of designated lands in their most natural condition
DATE: Enacted on September 3, 1964
SIGNIFICANCE: The U.S. Congress has designated almost 5 percent of the total U.S. land area as wilderness under the provisions of the Wilderness Act, which created the National Wilderness Preservation System. Designated wilderness areas are protected from development and from environmentally disruptive activities.

The 1964 Wilderness Act established the National Wilderness Preservation System (NWPS), gave the U.S. Congress authority to designate wilderness areas, and directed the secretaries of the interior and agriculture to review lands for possible wilderness designation. The act initially set aside 54 areas—a total of 3.6 million hectares (9 million acres) of federal Forest Service land—for wilderness classification. In 1968 Congress began adding wilderness areas. By 1999 there were 631 wilderness areas in forty-four states, totaling nearly 42.1 million hectares (104 million acres). A decade later, that total had grown to 756 wilderness areas (including a tropical rain forest in Puerto Rico, the first wilderness area designated in a U.S. territory) that all together occupied more than 44 million hectares (109 million acres). Wilderness areas are located within national forests, wildlife refuges, and parks and are managed by a host of agencies, including the U.S. Forest Service, the Bureau of Land Management (BLM), the National Park Service, and the Fish and Wildlife Service.

The land area of the United States totals 914 million hectares (2.3 billion acres); by 2010 approximately 4.7 percent of that had been designated wilderness. Just over half of this land—23.2 hectares (57.4 million acres)—is in the state of Alaska and accounts for roughly 16 percent of land in the state. Alaska is also the state in which the largest amount of land has actually been set aside: The Alaska National Interest Lands Conservation Act (ANILCA) of 1980 more than tripled the NWPS by establishing 35 new wilderness areas totaling more than 22.7 million hectares (56 million acres). Excluding Alaska, less than 3 percent of land in the United States is classified as wilderness or has been recommended for such designation.

The Wilderness Act defines wilderness as federal land "where the earth and its community of life are untrammeled by man, where man himself is a visitor who does not remain." Although numerous exceptions are made, the act generally prohibits commercial activities, motorized and mechanical access, permanent roads, and human-made structures and facilities within wilderness areas. An area may be determined to be suitable for wilderness designation if it is

> an area of undeveloped land retaining its primeval character and influence, without permanent improvements or human habitation, which is protected and managed so as to preserve its natural conditions and which (1) generally appears to have been affected primarily by the forces of nature, with the impact of man's works substantially unnoticeable; (2) has outstanding opportunities for solitude or primitive and unconfined type of recreation; (3) has at least five thousand acres of land or is of sufficient size as to make practicable its preservation and use in an unimpaired condition; and (4) may also contain ecological, geological, or other features of scientific, educational, scenic, or historic value.

Congress has the authority to designate areas as wilderness and uses its power to do so under the act. Designation is permanent, and new lands are added as Congress sees fit.

### PERMITTED AND PROHIBITED USES

Although wilderness areas are protected to preserve their natural conditions, a number of nonmotorized activities—such as horseback riding, hiking, camping, fishing, and hunting—are allowed in them. Preexisting and valid extractive uses are also allowed to continue until the permits granted for such activities expire, are abandoned, or are purchased by the government. Preexisting grazing is allowed to continue as long as it is consistent with sound resource management practices. In addition, the Wilderness Act honors all federal-state relationships with regard to state water laws and state fish and wildlife responsibilities. Activities that are generally not allowed in these areas include mining, timber harvesting, water development, mountain biking, and use of any motorized equipment such as snowmobiles and all-terrain vehicles. Allowances for some of these activities made in the act can be seen as a compromise between preservationists and those resource interests concerned with grazing, mining, timber harvesting, and water development.

Given the definition of wilderness and the amount of federally owned land in the western United States, congressionally classified wilderness is a particularly western phenomenon. Whereas states such as Alaska, Arizona, California, Idaho, and Washington have more than 1.6 million hectares (4 million acres) of wilderness each within their borders, the nonwestern states Connecticut, Delaware, Iowa, Kansas, Maryland, and Rhode Island have none.

Although Congress makes the final decision as to the suitability of additional areas for wilderness designation, land management agencies such as the Forest Service and the BLM are often responsible for making official recommendations. For example, the Federal Land Policy and Management Act of 1976 directed the BLM to review land it administers for possible wilderness designation, and the Forest Service has gone through two Roadless Area Review and Evaluation (RARE) plans and the 2001 Roadless Area Conservation Rule to determine suitable wilderness areas. Such plans have often been criticized for not designating enough land area and for refusing to designate land with great economic potential, thus leaving "rocks and ice" for wilderness classification and those lands with economic value under multiple-use management.

*Martin A. Nie*
*Updated by Karen N. Kähler*

FURTHER READING

Campaign for America's Wilderness. *People Protecting Wilderness for People: Celebrating 40 Years of the Wilderness Act*. Washington, D.C.: Author, 2004.

Dawson, Chad P., and John C. Hendee. *Wilderness Management: Stewardship and Protection of Resources and Values*. 4th ed. Boulder, Colo.: WILD Foundation, 2009.

Frome, Michael. *Battle for the Wilderness*. Rev. ed. Salt Lake City: University of Utah Press, 1997.

Harvey, Mark. "Loving the Wild in Postwar America." In *American Wilderness: A New History*, edited by Michael Lewis. New York: Oxford University Press, 2007.

Hays, Samuel P. *Wars in the Woods: The Rise of Ecological Forestry in America*. Pittsburgh: University of Pittsburgh Press, 2007.

Nash, Roderick. *Wilderness and the American Mind*. 4th ed. New Haven, Conn.: Yale University Press, 2001.

Scott, Doug. *The Enduring Wilderness: Protecting Our Natural Heritage Through the Wilderness Act*. Golden, Colo.: Fulcrum, 2004.

## Wilderness areas

CATEGORIES: Preservation and wilderness issues; land and land use

DEFINITION: Natural, undeveloped areas in the United States that are protected under the Wilderness Act of 1964

SIGNIFICANCE: The designation of large areas of land as protected wilderness areas is the subject of ongoing debate in the United States, with preservationists asserting that more areas need to be protected and critics arguing that the natural resources found in these areas should be available for use.

Preserving areas of unspoiled nature is a relatively new idea, and in the United States, this idea began to make sense to many Americans only when the seemingly inexhaustible wilderness of North America had been, in fact, nearly exhausted. In 1924, at the urging of Forest Service employee and influential conservationist Aldo Leopold, 305,500 hectares (755,000 acres) of the Gila National Forest in New Mexico were set aside as the first federally protected wilderness, the Gila Primitive Area. As the system of primitive areas grew, environmentalists became concerned about inconsistent management and about the fact that these areas were protected only by agency policy and not by law. They began lobbying for federal legislation that would designate and protect wilderness areas throughout the United States. The concept of preserving wilderness was strongly opposed, however, by many of those who made their livings by using natural resources; these included people involved in ranching and those in the timber and mining industries. They saw protected wilderness lands, which often had great economic value, as being "locked up" for the pleasure of a few.

### Legislation

On September 3, 1964, after eight years of debate and compromise, President Lyndon B. Johnson signed the Wilderness Act, creating the National Wilderness Preservation System (NWPS), which consisted of fifty-four areas totaling 3.6 million hectares (9 million acres). The act states:

> A wilderness, in contrast with those areas where man and his own works dominate the landscape, is hereby recognized as an area where the earth and its community of life are untrammeled by man, where man himself is a visitor who does not remain.

The act defines the mechanism for adding more areas to the system in the future. To be considered, an area must be at least 2,023 hectares (5,000 acres) "or of manageable size." This is a far cry from the early days of wilderness advocacy, when the minimum size was thought to be 202,000 hectares (500,000 acres), or, as Aldo Leopold put it, "large enough to absorb a two-week pack trip." Designated wildernesses become part of the NWPS. All roads, structures, and other installations are prohibited in designated wilderness, as is the use of motorized equipment or any mechanical transport. These areas of wild nature have been, and continue to be, the focus of intense controversy regarding their designation and management.

A significant addition to the NWPS came in 1975 with the passage of the Eastern Wilderness Act. The lack of pure, untouched wilderness in the eastern states led to the loosening of the strict standards of the original act to allow the inclusion of ecologically significant areas that show more impact from human activities than would originally have been permitted. In this way, sixteen areas totaling 83,770 hectares (207,000 acres), from 8,900-hectare (22,000-acre) Bradwell Bay in Florida to the 5,670-hectare (14,000-acre) Lye Brook Wilderness in Vermont, were added to the system. As of the late 1990's, the wilderness system encompassed more than 650 areas, ranging in size from the 2-hectare (5-acre) Oregon Islands Wilderness to the 3.6 million hectares (9 million acres) of the Wrangell-St. Elias Wilderness in Alaska, for a total of more than 40 million hectares (100 million acres). By 2010, in part as a result of the passage of the Omnibus Public Land Management Act of 2009, the number of wilderness areas had grown to 756, with a total of more than 44.1 million hectares (109 million acres). Of this total, 52 percent was in the state of Alaska.

### Debates and Controversies

Although the amount of protected land may seem quite large, preservationists point out that only about 5 percent of the landscape of the United States is protected in its natural state. Some large areas continue to be fought over, such as the fragile Arctic coastal plain of Alaska, home of vast caribou herds and underlain by large oil deposits. Idaho, which is among the top three states in the lower forty-eight in terms of wilderness land area (exceeded only by California and Arizona), still has millions of hectares of undeveloped roadless land that many believe should be protected. Wilderness advocates also point out that many wilderness areas, as well as national parks and other protected lands, have illogical political boundaries, unrecognized by grizzly bears and other important wildlife species. They argue that areas between and adjacent to designated wilderness areas should often be protected as well, to create units based on natural, ecological boundaries.

After a wilderness area is designated, the focus shifts to the maintenance of its desired qualities, leading to the paradox of "wilderness management." Although recreation is only one of the stated uses of wilderness—the others being scenic, scientific, educational, conservation, and historic—agency efforts and budgets are based primarily on the need to manage the often vast numbers of human visitors. One of the stated purposes of preserving wilderness areas is to provide for "primitive and unconfined recreation," but another consideration is the protection of the resource itself. At what point do the camping and trail restrictions, quotas, and permits needed to protect the resource impinge on the unconfined recreation of the visitor?

Another issue related to wilderness areas is that of wildfire suppression. It is now understood that fire is an important component of most ecosystems, but past policies of fire suppression have left unnatural fuel conditions in many areas. Should managers allow wildfires to burn, even though these fires are likely be larger and more destructive than natural, periodic fires of the past? Other major controversies center on the reintroduction of wildlife species (such as the wolf and the grizzly bear) to wilderness areas, the disposition of long-standing mining and drilling claims, and the flying of aircraft over, or even into, remote wilderness areas.

*Joseph W. Hinton*

FURTHER READING

Allin, Craig W. *The Politics of Wilderness Preservation*. 1982. Reprint. Fairbanks: University of Alaska Press, 2008.

Dawson, Chad P., and John C. Hendee. *Wilderness Management: Stewardship and Protection of Resources and Values*. 4th ed. Boulder, Colo.: WILD Foundation, 2009.

Frome, Michael. *Battle for the Wilderness*. Rev. ed. Salt Lake City: University of Utah Press, 1997.

Nash, Roderick. *Wilderness and the American Mind*. 4th ed. New Haven, Conn.: Yale University Press, 2001.

Scott, Doug. *The Enduring Wilderness: Protecting Our Natural Heritage Through the Wilderness Act*. Golden, Colo.: Fulcrum, 2004.

## Wilderness Society

CATEGORIES: Organizations and agencies; activism and advocacy; preservation and wilderness issues

IDENTIFICATION: American nonprofit organization dedicated to the protection and preservation of wilderness areas and wildlife

DATE: Established in 1935

SIGNIFICANCE: The Wilderness Society has been instrumental in the passage of major conservation legislation in the United States, including the Wilderness Act of 1964, the Wild and Scenic Rivers and National Trails System Acts of 1968, and the National Forest Management Act of 1976.

In January, 1935, a group of eight dedicated conservationists organized the Wilderness Society in Washington, D.C. Among the participants were Robert S. Yard, publicist for the National Park Service; Benton MacKaye, known as the "father of the Appalachian Trail"; Robert Marshall, head of recreation and lands for the U.S. Forest Service; and Aldo Leopold, a wildlife ecologist at the University of Wisconsin. Leopold believed that the new society would form a cornerstone for efforts to preserve America's vanishing wilderness.

After much dedicated work and pressure by the Wilderness Society, the Wilderness Act was finally signed into law by President Lyndon B. Johnson on September 3, 1964. This act established the National Wilderness Preservation System (NWPS), which enabled the U.S. Congress to set aside selected areas within national forests, national parks, national wildlife refuges, and other federal lands as units to be kept permanently unchanged by humans. There would be no roads, structures, vehicles, or any significant impacts of any kind in these selected areas. The Wilderness Act initially designated approximately 3.6 million hectares (9 million acres) as wilderness.

The Wilderness Society has since had a hand in the passage of several other major public lands bills, including the Wild and Scenic Rivers Act and the National Trails System Act, both passed in 1968, and the National Forest Management Act of 1976. The organization's efforts have helped to contribute a total of 44 million hectares (109 million acres) to the NWPS. In particular, the Alaska National Interest Lands Conservation Act of 1980 designated 22.7 million hectares (56 million acres) of pristine land for protection, and the California Desert Protection Act of 1994 designated 3.2 million hectares (8 million acres) of desert lands. After a ten-year effort, the Wilderness Society was instrumental in the enactment of the National Wildlife Refuge System Improvement Act of 1997, which strengthened protections for wildlife in all national wildlife refuges.

During the 1980's and 1990's, despite the successes of the Wilderness Society, public lands continued to be compromised and degraded by air and water pollution, excessive development, road building, logging, cattle grazing, mining, and recreational activities. By the late 1990's, the Wilderness Society had focused on the preservation of a number of high-priority areas, with the overall goal of creating a nationwide network of wildlands. Key campaigns were launched in Montana, California, Idaho, Nevada, Utah, Colorado, Texas, Vermont, the Pacific Northwest, and the southern Appalachians.

In 1997, with backing from the Wilderness Society,

the U.S. Congress blocked a plan to create a massive gold mine just outside Yellowstone National Park by appropriating money from the Land and Water Conservation Fund to purchase the mining claims. Similarly, the society helped create public pressure that led the Du Pont Corporation to defer plans for a titanium mine on the border of the Okefenokee National Wildlife Refuge in Georgia. Working closely with local and national groups during 1996 and 1997, the society convinced the federal government to withdraw a proposed logging plan for nine national forests in the Sierra Nevada. Additionally, in response to a Wilderness Society lawsuit, a federal judge blocked logging in four national forests in Texas, pointing out that it would be detrimental to wildlife habitats.

Despite the loosening of some environmental protections under the presidential administration of George W. Bush, the Wilderness Society enjoyed several notable victories during this period. Among them were the 2001 adoption by the U.S. Forest Service of the Roadless Area Conservation Rule, which protects roughly 24 million hectares (60 million acres) within national forests, and the 2005 addition of the first tropical rain forest—4,047 hectares (10,000 acres) in Puerto Rico—to the U.S. national forest system. In late 2008 Wilderness Society staff and policy experts met with U.S. president-elect Barack Obama's transition team and urged that the new administration take action to address climate change and protect roadless forests, the Arctic National Wildlife Refuge, and other fragile lands threatened by oil and gas drilling. The 2009 Omnibus Public Land Management Act, passed during the first year of Obama's presidency, protects 0.85 million hectares (2.1 million acres) of new wilderness areas in nine states.

*Alvin K. Benson*
*Updated by Karen N. Kähler*

### Further Reading

Kline, Benjamin. *First Along the River: A Brief History of the United States Environmental Movement.* 3d ed. Lanham, Md.: Rowman & Littlefield, 2007.

Maher, Neil M. "The Great Conservation Debate." In *Nature's New Deal: The Civilian Conservation Corps and the Roots of the American Environmental Movement.* New York: Oxford University Press, 2008.

Sutter, Paul. "New Deal Conservation: A View from the Wilderness." In *FDR and the Environment*, edited by Henry L. Henderson and David B. Woolner. New York: Palgrave Macmillan, 2005.

## Wildfires

- **Categories:** Preservation and wilderness issues; resources and resource management
- **Definition:** Fires occurring in wilderness or open country in various vegetation types, generally characterized by large size, rapid flame spread, intensity of heat and smoke, and difficulty of prediction or control of behavior
- **Significance:** Growing human control of fire has resulted in intensification of the human relationship with, and responsibility for, fire in the environment. This growing control has resulted in ongoing alterations in the definition of "wild" fire. These changes in definition are themselves reflective of fire-mediated alteration and modification of the preexisting environment by humans.

For more than 450 million years, all fires in plant matter on the earth were both "natural" and "wild." Over the past half million to a million years, and particularly during the past 20,000 years, the "control" (meaning both use and suppression) of fire by humans has changed not only our understanding of "natural fire" and "wildfire" but also the character of myriad environments and the human cultures sustained by them. From the effects of slash-and-burn agriculture on species diversity to the effects of global climate change on future human urbanization patterns, human control of fire has had tremendous consequences—even, paradoxically, increasing the number, intensity, and duration of wildfire events.

### Use and Suppression of Fire

Although wildfires occur on every continent except Antarctica, the consequences of the use and the suppression of fire are best seen in four regions: Amazonia, equatorial Africa, Australia, and western North America. An understanding of the dynamics of use and suppression in regard to wildfire requires expansion of the traditional "fire triangle" of ignition source, combustible material, and oxygen into a "wildfire hexagon" through the addition of three key factors: topography (the "shape" of the landscape), climate or weather (particularly in regard to the effects of drought on fuel moisture and strong winds on fire behavior), and human interaction (both in shorter-term responses to individual fire events and in longer-term transformation of fire-prone environments).

Weather, topography, and the presence of atmospheric oxygen are not easily altered on large scales within short time frames, so only three elements of the wildfire hexagon—ignition source, combustible material, and human interaction—can be readily affected by short-term human activity. Because fire has long been a tool useful in clearing the land for subsequent agricultural production, in many areas of Amazonia and equatorial Africa people intervene by enhancing natural fire cycles through purposely introducing ignition sources and combustible material into the environment. Because fire has also been perceived as a threat to valuable timber, homes, and other property, however, in many areas of Australia and western North America people intervene by suppressing natural fire cycles through reducing ignition sources where possible or, once a fire has been ignited, attempting to reduce the combustibility or availability of fuels through firefighting activities.

Problems with these strategies of use and suppression arise when the use becomes too careless or the suppression too careful. Fire used to clear land in Amazonia and equatorial Africa often ends up making its way into surrounding rain forest, there becoming wildfire. Years and decades of too-careful fire suppression in western North America and Australia have resulted in thickety forests so overburdened with ladder fuels—leading from flashy fuels such as grass and duff along the ground through a midlevel of shrubs and younger trees, finally to the topmost canopy of mature trees—that once a fire gets started in such forest it is much more prone to become a catastrophically destructive burn. Although "natural" and "controlled" are usually seen as opposites, human control of fire in both the "use" and "suppression" situations outlined above has separated "natural" from "wild" such that the resulting wildfires are both uncontrolled and unnatural.

Patchwork use of broadcast burning in wildland—by hunters to stampede game, by pastoralists to open up grazing lands, and by agriculturalists to prepare land for planting—arguably originated many tens of thousands of years ago. Systematic, widespread suppression of fire in wildland arguably began only in response to the Great Fire of 1910 in the United States. With the continued unprecedented growth in human population over the past three centuries, however, the increasing number of wildfires stemming from both fire use and fire suppression has resulted in many of the same negative effects on the environment, from the physical (increased erosion, landslides, mudflows, flash flooding, and altered water quality) to the biological (loss of species habitat, introduction of invasive species, and declines in biodiversity).

Increases in population and demographic changes have meant that more and more land that was formerly wild has become increasingly bordered and penetrated by housing tracts along the wildland-urban interface (WUI; also known as the wildland-urban intermix). The upshot has been still more property at risk, with consequent pressures to suppress fire more thoroughly in naturally fire-prone but increasingly populated and economically valuable areas—with the result that more fuels build up year upon year until, when ignition finally comes, the result is too often a devastating wildfire.

Wildfires affect not only physical, biological, and economic aspects of landscapes and watersheds but the atmosphere as well. Smoke, soot, ash, ozone, greenhouse gases (such as carbon dioxide), and other fire by-products are lofted by wildfire not only throughout the troposphere (the lowest layer of the atmosphere) but also as high as the lower stratosphere, where their influence on human health and on climate can be global.

## Striking a New Balance

Growing awareness that too-careless use and too-careful suppression of fire in wildlands have resulted in increases in the number, duration, and intensity of wildfires—with cascading local, regional, and global ecological and economic effects—has led fire experts to reevaluate approaches to wildfires. This reevaluation is based in an increasingly nuanced ecological understanding of the roles of natural fire, fire cycles, weather patterns, and fire-dependent, fire-tolerant, and fire-intolerant plant adaptations.

The recognition that many natural landscapes contain plants that are fire-dependent or fire-tolerant (not only in western North American and Australian wildlands but also in Southeast Asia and South Africa) has called into question policies of too-careful or complete fire suppression, particularly when such policies have resulted in the increased presence of both flashy, fire-intolerant vegetation and higher fuel loads generally. Because lower-intensity fires can reduce fuel loads and reduce or eliminate often invasive fire-intolerant vegetation, fire has been selectively allowed back into these wildlands. Carefully monitored against escape onto higher-value locations, fires that have been naturally caused (most often by lightning, but also at times by volcanic eruptions or meteor strikes) are allowed to burn so that they can fulfill their ecological role.

Controlled burning (generally referred to as prescribed burning in the United States) is a fire management strategy in which wildland fires are purposely ignited under favorable weather conditions, with the goal of creating lower-intensity fires that help clean out accumulated fuels, foster higher levels of species diversity, and reduce the future risks of intense, long-lived wildfires. The goal of such "allowed fire" strategies is to emulate natural fire through the use of controlled fire of the right type, in the right place, and at the right time.

Just as too-careful fire suppression has had to give way to some controlled use of fire in an attempt to emulate natural fire, so too-careless fire use has had to be curbed through prevention of ignition—again in an attempt to emulate natural fire. Because slash-and-burn agricultural practices damage fire-resistant rain forests and encourage the growth of flammable brush (and the more frequent occurrence of future wildfires), those wishing to curtail careless fire use in rain-forest areas—whether neighboring landowners with flammable tree crops and orchards or environmentalists wishing to preserve rain-forest species diversity—have had to exert pressure on slash-and-burn agriculturalists to be more circumspect in igniting, more careful in monitoring, and more thorough in suppressing fires lit to clear land.

Because human interaction can change both the fuel and the ignition factors in the wildfire hexagon, the fastest and most powerful way to affect the risk of future wildfire is to change the way humans interact with their environments. Those who in the past suppressed fire in order to further their interests now find they must at times incorporate fire use in their approach, and those who in the past used fire to further their own interests now find they must also at times incorporate fire suppression in their approach.

*Howard V. Hendrix*

### Further Reading

Carle, David. *Introduction to Fire in California*. Berkeley: University of California Press, 2008.

Egan, Timothy. *The Big Burn: Teddy Roosevelt and the Fire That Saved America*. Boston: Houghton Mifflin Harcourt, 2009.

Holbrook, Stewart H. *Burning an Empire: The Study of American Forest Fires*. 1943. Reprint. New York: Macmillan, 1960.

United Nations Food and Agriculture Organization. *FAO Meeting on Public Policies Affecting Forest Fires*. Rome: Author, 1999.

Wilson, Bill, et al., eds. *Forest Policy: International Case Studies*. Wallingford, England: CABI, 1998.

## Wildlife refuges

CATEGORIES: Animals and endangered species; land and land use

DEFINITION: Regions of land or water set aside by governments or private organizations to protect and preserve one or more species of wildlife

SIGNIFICANCE: The U.S. National Wildlife Refuge System has endured congressional debate and public scrutiny involving environmental issues related to the societal, governmental, and commercial use of designated sanctuaries, culminating in the 1997 National Wildlife Refuge System Improvement Act and the transformation of America's refuges into multiple-use systems.

Prior to 1900, the U.S. federal government aggressively raised much-needed revenue and rewarded growing commerce by selling or giving away nearly 405 million hectares (1 billion acres) of land to states, homesteaders, veterans, railroads, and businesses. President Theodore Roosevelt initiated the protection of habitat for wildlife in 1903 when he set aside Pelican Island, a 1.2-hectare (3-acre) ecosystem of barren sand and scrub in Florida's Indian River, as a federal reservation to protect birds from hunters supplying plumes to the fashion industry. Inspired while camping in California's Yosemite Valley with naturalist John Muir, Roosevelt established more than fifty wildlife refuges, five national parks, and eighteen national monuments, such as the Grand Canyon. He also greatly increased the area of lands designated as national forests before leaving his second term in office.

To preserve additional lands "for our children and their children's children forever, with their majestic beauty all unmarred," Roosevelt guaranteed land for future refuges by separating other federal public domain regions such as national forests and rangelands

## Milestones in Wildlife Protection

| YEAR | EVENT |
|---|---|
| 1870 | The first state wildlife refuge is established in California. |
| 1898 | Kruger National Park is established in South Africa for the preservation of big game. |
| 1900 | The Lacey Act regulates the interstate commerce of birds and mammals; the act is supplemented by a similar act for black bass in 1926. |
| 1903 | The first federal bird sanctuary is established by President Theodore Roosevelt at Florida's Pelican Island. |
| 1908 | Theodore Roosevelt calls a conference of state governors and related officials to inventory natural resources in the United States. |
| 1916 | The National Park Service is established and forbids hunting within its jurisdiction. |
| 1929 | The Migratory Bird Conservation Act provides for a system of refuges along major flyways. |
| 1934 | The Duck Stamp Act requires hunters of migratory fowl to purchase duck stamps with their waterfowl licenses; proceeds are used to establish wildlife refuges. |
| 1937 | Taxes on arms and ammunition are used for wildlife preservation. |
| 1940 | The National Wildlife Refuge System is established by a consolidation of the Bureau of Biological Survey and the Bureau of Fisheries; its mission includes biological research and administration as well as enforcement of federal legislation. |
| 1966 | The National Wildlife Refuge System Administration Act mandates that all refuge uses be compatible with the primary purpose for which the refuge was established. |
| 1970 | The National Environmental Policy Act, as well as other legislation designed to combat pollution, is passed. |
| 1973 | The Endangered Species Act, which updates prior acts in 1966 and 1969, requires refuge managers to protect certain species of flora and fauna. |
| 1980 | The Alaska National Interest Lands Conservation Act doubles the amount of land in the U.S. refuge and park systems. |
| 1987 | A federal court rules that the U.S. Fish and Wildlife Service is responsible for policing the existing ban on spring hunting by native groups in Alaska. |
| 1992 | Research commissioned by Defenders of Wildlife finds that the National Wildlife Refuge System is grossly inadequate. |
| 1997 | The National Wildlife Refuge System Improvement Act establishes a revamped multiple-use mission statement for refuge habitat conservation. |
| 2001 | The Roadless Area Conservation Rule, a federal policy initiative designed to protect national forests from commercial development, is issued. |
| 2005 | More than 4,000 hectares (approximately 10,000 acres) of Puerto Rican rain-forest land are added to the U.S. national forest system. |
| 2008 | Leaders of the Wilderness Society urge U.S. president-elect Barack Obama to address climate change and increase protection of national refuge lands. |
| 2009 | The Omnibus Public Land Management Act adds 850,000 hectares (2.1 million acres) of new wilderness areas in nine U.S. states. |

from the control of commercial interests. More than 90 percent of the refuge land area in existence in the United States in the early years of the twenty-first century resulted from Roosevelt's foresight, which enabled the National Wildlife Refuge System to grow larger than the national park system and entail nearly 4 percent of the surface area of the United States.

With vital assistance by private individuals and organizations such as the Nature Conservancy and the National Audubon Society, wildlife refuges have been established for waterfowl, big game, small resident game, and colonial nongame birds. Wildfowl refuges, easily the most plentiful, are geographically patterned to supply breeding, wintering, resting, and feeding areas along the four major North American migration flyways. The sportsmen who were essential in establishing many national refuges ensured that hunting would be permitted on most sanctuary lands, with trapping allowed on many. Although the entire National Wildlife Refuge System logs substantial numbers of hunting visits annually, visitors who are interested in wildlife education and photography outnumber hunters and anglers by more than four to one.

The National Wildlife Refuge System—overseen by the U.S. Fish and Wildlife Service, which is part of the Department of the Interior—is the most comprehensive nature protection network in the world. The entire system includes more than 60.7 million hectares (150 million acres) inside refuge boundaries, encompassing 553 national wildlife refuges and other units as well as 38 wetland management districts. Nearly all of the refuges in the system are open to the public. Endangered and threatened species are supported on almost 60 refuges created specifically for that purpose, and many urban refuges have been established near large cities. At least one wildlife refuge has been established in each of the fifty states, and several are found in overseas possessions from Puerto Rico to American Samoa in the South Pacific. The two largest refuges are in Alaska: The Arctic National Wildlife Refuge and the Yukon Delta National Wildlife Refuge are both more than 7.7 million hectares (19 million acres) in size. The smallest refuge in the system is the Mille Lacs National Wildlife Refuge in Minnesota, which is just 0.24 hectare (0.60 acre) in size.

### Environmental Management Issues

Creating and maintaining a refuge to provide food, cover, and protection from human development for wildlife is considerably more difficult than simply sequestering an area and allowing nature to run its course; continual management of resources is imperative to keep delicate ecosystems in balance. The many tasks that humans must perform on refuge lands include fixing broken floodgates and cleaning clogged ditches, seeding wildlife foods, plowing and burning areas that have been overrun with unwanted vegetation, and closing off areas from the public during sensitive periods for animals, such as mating and birthing seasons. In addition, refuge managers must often battle for their lands' shares of rapidly declining water supplies.

Although many refuges include areas just as spectacular as those within the national park system, the National Wildlife Refuge System as a whole was for many years not well utilized by the public. The U.S. Congress, observing this lack of public use, leased some of these public lands for commercial purposes such as grazing, farming, oil drilling, mining, logging of timber, military maneuvers, and motorized recreation. However, as public use of refuges increased during the 1980's and 1990's, Congress added eighty new refuges to the system, creating such a backlog of environmental preservation issues that the legislators then began to contemplate selling some areas to pay for maintenance. With minimal budgets, refuge managers are charged with making certain that all activities that take place within their refuges are compatible with wildlife while still allowing potentially destructive activities such as off-road driving and motorcycling, powerboating, and commercial fishing. The maintenance of biodiversity is also an important goal of managers; the wildlife refuge system provides homes for about 700 species of birds, 220 species of mammals, 250 species of reptiles and amphibians, and 200 species of fish, in addition to innumerable species of plants.

Refuge areas owned by private corporations have often sacrificed key habitat for short-term economic gain with little regard for long-term environmental and social consequences. However, business executives have realized that environmental issues are of genuine concern to most Americans. In response to increased environmental awareness and pressure from consumers, employees, and stockholders, many large businesses have implemented stewardship strategies that protect natural resources, enhance wildlife habitat, and provide for public enjoyment of their underdeveloped land.

## Key Legislative Actions

Following Theodore Roosevelt's initial work, new additions to the refuge system came slowly until the Dust Bowl years of the 1930's, when migratory bird populations became depleted. Congress then passed the 1934 Migratory Bird Hunting and Conservation Stamp Act, widely known as the Duck Stamp Act, which added a conservation fee to the price paid for every waterfowl license purchased by a hunter; the revenue collected in this way enabled the Fish and Wildlife Service to acquire wetlands along major bird migration flyways. Additional moneys to purchase refuge lands came from the Land and Water Conservation Fund set up in the 1960's to increase public space for outdoor recreation, which generated considerable revenues from offshore oil drilling leases. Passage of the 1980 Alaska National Interest Lands Conservation Act (ANILCA) enabled both the refuge system and the national park system to double in size. Although 96 percent of all refuge units are outside Alaska, Alaska contains about 83 percent of National Wildlife Refuge System lands.

The 1973 Endangered Species Act spurred managers of refuges to make more concessions for certain species of flora and fauna. At the beginning of the twenty-first century, refuges in the U.S. system harbored about 280 threatened or endangered species. Surveys in the late 1980's by the Fish and Wildlife Service and the General Accounting Office revealed that more than 60 percent of refuges were permitting activities known to be harmful to wildlife. The most harmful practices, such as military activities and drilling, were not under the control of the Fish and Wildlife Service. The high-profile activist group Defenders of Wildlife organized a citizens' commission in 1992 that confirmed that the National Wildlife Refuge System was "falling far short of meeting the urgent habitat needs of the nation's wildlife" and was suffering from "chronic fiscal starvation and administrative neglect."

The National Wildlife Refuge System Improvement Act, signed into law by President Bill Clinton on October 9, 1997, dramatically shifted the priorities of the refuge system from its original sole purpose of protecting wildlife to the formation of a multiple-use system. The legislation redefined the system's mission regarding the conservation of habitat for fish, wildlife, and plants; designated priority public uses such as hunting, fishing, wildlife observation and photography, and environmental education and interpretation; and required that the environmental health of the refuge system be maintained. This monumental bill gave hunting, fishing, commercial trapping, and recreation equal status in the refuges with the conservation of plants, birds, and animals. It also limited new or secondary refuge use to activities compatible with wildlife protection and made legislative changes more difficult for future congressional cycles. The principle of multiple use, however, continues to allow and may result in increased mining, drilling, grazing, logging, and motorized recreation on refuge lands, in addition to increased military training, including bombing and tank and troop exercises. Upon signing the bill, Clinton stated that he "hoped and trusted that the process by which this bill was enacted will serve as a model for future congressional action on other environmental issues," with the future of the National Wildlife Refuge System to be shaped by the future of other bills such as the Clean Water Act, the Wetlands Protection Act, and the Endangered Species Act.

*Daniel G. Graetzer*

## Further Reading

Butcher, Russell D. *America's National Wildlife Refuges: A Complete Guide.* 2d ed. Lanham, Md.: Taylor Trade, 2008.

Dolin, Eric Jay. *Smithsonian Book of National Wildlife Refuges.* Washington, D.C.: Smithsonian Institution Press, 2003.

Nelson, Lisa. "Wildlife Policy." In *Western Public Lands and Environmental Politics,* edited by Charles Davis. 2d ed. Boulder, Colo.: Westview Press, 2001.

Patent, Dorothy Hinshaw. *Places of Refuge: Our National Wildlife Refuge System.* Boston: Houghton Mifflin, 1992.

Riley, Laura, and William Riley. *Guide to the National Wildlife Refuges.* Rev. ed. New York: Macmillan, 1992.

# World Heritage Convention

CATEGORIES: Treaties, laws, and court cases; preservation and wilderness issues

THE CONVENTION: International agreement to protect designated sites of great cultural, historic, or natural value

DATE: Opened for signature on November 16, 1972

SIGNIFICANCE: The World Heritage Convention promotes protection of the environment by obligating signatory nations to identify, maintain, and preserve important natural and cultural sites within their territories as part of the universal heritage of humanity.

The United States proposed the World Heritage Convention in 1972 to commemorate the one hundredth anniversary of the establishment of Yellowstone National Park and was the first nation to sign the convention when it was adopted by the United Nations Educational, Scientific, and Cultural Organization (UNESCO). The convention—formally titled the Convention Concerning the Protection of the World Cultural and National Heritage—essentially promotes the U.S. national park concept worldwide. By the time of the convention's twenty-fifth anniversary in 1997, nearly 150 nations had ratified the agreement and had placed more than five hundred sites on the World Heritage List. By the end of the first decade of the twenty-first century, 187 nations had ratified the convention, and more than nine hundred cultural and natural sites had been named.

Nations that are signatories to the World Heritage Convention nominate sites within their own borders for inclusion on the World Heritage List. The nominations are reviewed by the World Heritage Committee, a body consisting of the representatives of twenty-one signatory nations; the representatives are elected by the General Assembly of the signatory nations. The committee also places sites threatened by natural disaster or civil strife on the List of World Heritage in Danger.

In signing the World Heritage Convention, nations pledge to identify, maintain, and preserve important natural and cultural sites within their territories as part of the universal heritage of humanity; they also pledge to promote and publicize these sites for worldwide public enlightenment. In addition, member nations assist one another with studies, advice, training, and equipment necessary to resolve problems, restore damaged areas, and establish programs to protect, preserve, and publicize the sites. The World Heritage Committee offers technical advice and monetary assistance through its World Heritage Fund. Individual nations also offer direct nation-to-nation assistance.

To be included on the World Heritage List, sites must possess outstanding, universally recognized cultural or natural features. Sites include both human-made constructions and natural areas. Selected natural sites include areas that represent major stages in the earth's evolutionary history, ongoing geological processes or biological evolution, or human interaction with the environment; that contain unique, rare, or superlative national phenomena; or that provide habitats for rare or endangered plants and animals.

Each signatory nation maintains sovereignty over its sites and is responsible for site maintenance and protection. Listed sites in a given country include those owned by the national government (such as national parks and national historic landmarks) as well as those owned by state or tribal governments, local governments, and private groups or individuals, with the owners pledging to protect their properties in perpetuity.

Among the sites included on the World Heritage

## World Heritage Site Distribution, 2010

| Region | Nations with Sites | Types of Sites | | | | Share of World Sites (%) |
| --- | --- | --- | --- | --- | --- | --- |
| | | Cultural | Natural | Mixed | Total | |
| Africa | 30 | 42 | 32 | 4 | 78 | 9 |
| Arab States | 15 | 61 | 4 | 1 | 66 | 7 |
| Asia and Pacific | 31 | 138 | 51 | 9 | 198 | 22 |
| Europe and North America | 50 | 377 | 58 | 10 | 445 | 49 |
| Latin America and Caribbean | 25 | 86 | 35 | 3 | 124 | 14 |
| Total | 151 | 704 | 180 | 27 | 911 | 100 |

Source: United Nations Educational, Scientific, and Cultural Organization.
Note: Uvs Nuur Basin, which straddles Europe and Asia in Mongolia and Russia, is counted here as part of Asia.

List are the Great Wall of China, the Taj Mahal in India, Ecuador's Galápagos Islands, the Tower of London, and the massive Spanish fortifications at San Juan, Puerto Rico. North American sites include twenty in the United States; among these are Grand Canyon National Park in Arizona, Everglades National Park in Florida, Independence Hall in Philadelphia, Cahokia Mounds State Historic Site in Illinois, Pueblo de Taos in New Mexico, and the Statue of Liberty in New York.

*Gordon Neal Diem*

FURTHER READING

Di Giovine, Michael A. *The Heritage-scape: UNESCO, World Heritage, and Tourism.* Lanham, Md.: Lexington Books, 2009.

Leask, Anna, and Alan Fyall, eds. *Managing World Heritage Sites.* Burlington, Mass.: Butterworth-Heinemann, 2006.

McHugh, Lois. "World Heritage Convention and U.S. National Parks." In *American National Parks: Current Issues and Developments,* edited by Rony Mateo. Hauppauge, N.Y.: Novinka Books, 2004.


# World Summit on Sustainable Development

CATEGORY: Resources and resource management

THE EVENT: International conference devoted to finding ways to bring together developed and developing nations in the pursuit of sustainable and equitable development

DATES: August 26-September 4, 2002

SIGNIFICANCE: The World Summit on Sustainable Development brought together thousands of participants from all parts of the world and from all levels of society, and in doing so focused worldwide attention on the needs of developing nations and the impacts of development on the environment. The summit produce the Johannesburg Declaration on Sustainable Development, a statement of principles for international action and debate.

The 2002 World Summit on Sustainable Development was held in Johannesburg, South Africa, and is thus sometimes referred to as the Johannesburg Summit; it is also known as Earth Summit 2002. The purpose of the summit was in part to follow up on and discuss the progress made since the 1992 United Nations Conference on Environment and Development, known as the Earth Summit, which was held in Rio de Janeiro, Brazil. The 2002 summit also sought to focus the world's attention on the problematic effects of current methods of development for the planet's environmental health and the well-being of its peoples.

The charge of the summit was to bring into communication and collaboration a wealth of diverse powers, perspectives, and interests so that strategies could be devised and direct action taken toward a concrete plan for sustainable development. Participants addressed the challenges associated with a globalizing economy that increasingly suffers from mounting industrialization, urbanization, ecological degradation, and the depletion of natural resources, problems exacerbated by the world's growing population and ever-increasing demands for food, water, shelter, energy, health and sanitation services, and economic security.

At the Rio meeting in 1992, the international community adopted Agenda 21, an unprecedented global action plan for sustainable development that was founded on the conviction that human beings had reached a defining moment in the history of humankind and of the earth. Agenda 21 calls for a radical departure from the development policies of the past, which have not only increasingly devastated the environment but also contributed to gaping economic divisions between haves and have-nots around the globe, with consequent increases in the levels of poverty, hunger, disease, and illiteracy worldwide. Agenda 21 is intended to chart a new course to improve the living standards of the masses of hopeless impoverished at the base of the economic system and to restore the earth's ecological health in order to bring about a more prosperous future for humankind and for the planet. The 2002 summit provided a fresh opportunity for international leaders and world citizens to return to the plan laid out in Agenda 21—to assess progress toward its implementation, identify shortcomings in that progress, and renew commitment to quantified targets for its fuller implementation.

PLANNING AND PARTICIPANTS

The World Summit on Sustainable Development was planned and organized by a committee of the tenth session of the United Nations Commission on

Sustainable Development in four preparatory meetings held in 2001 and 2002. A bureau consisting of ten representatives, two from each of five world regions, steered the preparations for the summit and worked to raise international awareness of and support for the summit among world governments and lay groups. A key element of the summit planning was the United Nations Secretary-General's Advisory Panel, a team charged with exploring the challenges of sustainable development and making recommendations to the secretary-general concerning how those challenges might be addressed at the summit.

Because the summit was to be a mammoth global event and certainly the largest gathering of international delegates ever convened in Africa, its coordinators were committed to carrying out all aspects of the event according to the ecologically sound "best practices" that the summit would be promoting. The Greening the WSSD Initiative was established to oversee the environmental impacts of the summit and to ensure that minimal waste would be generated by the thousands of participants.

During the summit, while international governmental representatives were meeting at the Sandton Convention Centre on the outskirts of Johannesburg, a nongovernmental forum was also under way at the nearby NASREC Expo Centre. Numerous side events coordinated by the United Nations, as well as parallel events coordinated by independent groups, ensured the broad participation and inclusiveness that the summit's planners considered to be key to the successful setting of realistic but high-reaching goals and gaining widespread commitment to those goals.

More than twenty thousand of people from every corner of the world and every level of society took part in various aspects of the summit, from heads of state and other national delegates to leaders of nongovernmental organizations, leaders in business and industry, and a broad variety of workers and trade unionists, farmers, indigenous peoples, local authorities, and members of the scientific and technological communities. The broad diversity of participants was intended to reflect the major groups identified in Agenda 21.

## Approaches and Outcomes

Recognizing that poverty in the developing world is a complex, multifaceted problem that is integrally related to nations' external indebtedness, internal corruption, global trade policies, lack of development capital, and many other factors, both national and international, participants in the World Summit on Sustainable Development sought to develop both global and country-specific plans for creating sustainable patterns of consumption and production. These included devising new tactics for optimizing global resource use, minimizing waste, transferring environmentally sound technologies across the globe, achieving greater equity in income distribution, and developing human resources in all corners of the planet through the generation of employment, the extension of basic education and professional training opportunities, and the establishment of effective primary and maternal health care systems. Whether the proposed sustainability initiatives were global or national, the summit encouraged a community-driven approach that would empower local and community groups and include women, youth, and children.

For the most part, the summit was deemed to be successful, because its broad participant base and the global media coverage it received focused international attention squarely on the problem of sustainability. Many celebrated the summit's progress regarding aid to developing nations. Critics, however, argued that the summit's success was reduced by the conspicuous absence of U.S. president George W. Bush from among the more than one hundred world leaders who gathered to address such crucial issues as the spread of acquired immunodeficiency syndrome (AIDS) and other diseases, the depletion of fish stocks around the world, and the need to promote environmentally friendly agriculture. Critics asserted that Bush's absence reduced negotiators' ability to address a number of tough issues, such as the problem of rich nations' policies of farm subsidies, which render poor nations incapable of competing with rich nations agriculturally. Many delegates also complained that the United States led the developed nations in resisting the setting of new targets to phase out export subsidies and other trade-distorting domestic supports.

*Wendy C. Hamblet*

### Further Reading

Bigg, Tom. "The World Summit on Sustainable Development: Was It Worthwhile?" In *Survival for a Small Planet: The Sustainable Development Agenda*, edited by Tom Bigg. Sterling, Va.: Earthscan, 2004.

Hens, Luc, and Bhaskar Nath. *The World Summit on*

Sustainable Development: The Johannesburg Conference. New York: Springer, 2005.

Speth, James Gustave, and Peter M. Haas. "From Stockholm to Johannesburg: First Attempt at Global Environmental Governance." In *Global Environmental Governance*. Washington, D.C.: Island Press, 2006.

Strachan, Janet R., et al. *The Plain Language Guide to the World Summit on Sustainable Development*. Sterling, Va.: Earthscan, 2005.

## Yellowstone National Park

CATEGORIES: Places; preservation and wilderness issues

IDENTIFICATION: U.S. national park located in Wyoming and parts of Montana and Idaho

SIGNIFICANCE: Yellowstone National Park is among the areas of the continental United States containing the widest biodiversity of both plant and animal species. Like all national parks, it faces the constant challenge of balancing protection of the ecosystems it encompasses with enabling access by people who want to take advantage of the land's scenic beauty and the recreational opportunities it offers.

Yellowstone National Park was the first national park, both in the United States and in the world. In 1806 John Colter became the first non-Native American to explore the region in which Yellowstone is located. In 1870 Henry Washburn, Nathaniel Langford, and Gustavus Doane spent more than a month investigating the area and documenting their expedition. Many of those who traveled through the region were deeply impressed by its beauty and spoke out in support of its being declared a public park, and in 1872 President Ulysses S. Grant signed the legislation creating Yellowstone National Park. Langford was appointed the first superintendent of the park; he served for five years and then was replaced by Philetus Norris, who secured the funding necessary to build the first rudimentary roads in the park.

Yellowstone National Park is approximately 8,990 square kilometers (3,470 square miles) in size and expands across the northwest corner of Wyoming as well as parts of Montana and Idaho. The park is home to black bears, grizzly bears, bison, coyotes, deer, elk, and moose that wander freely through the meadows; bobcats, mountain lions, and bighorn sheep live in the mountainous areas of the park. The gray wolf population that originally lived in the park was decimated by predator-control measures in the early twentieth century, but the species was reintroduced in 1995, and by 2010 more than seventeen hundred gray wolves were living within the park's protected habitat. In addition to the wide variety of wildlife, more than seventeen hundred species of plants and trees are found within the park's boundaries.

Yellowstone also contains numerous geological natural wonders. These include almost three hundred waterfalls, one of the world's largest volcanic calderas, and more than ten thousand geothermal features. Perhaps most famous among the geothermal features are the park's hot springs and geysers. Old Faithful Geyser erupts roughly every ninety minutes, much to the delight of the park's visitors. It reaches more than 93 degrees Celsius (200 degrees Fahrenheit) in temperature, releases thousands of gallons of water, and shoots water more than 55 meters (180

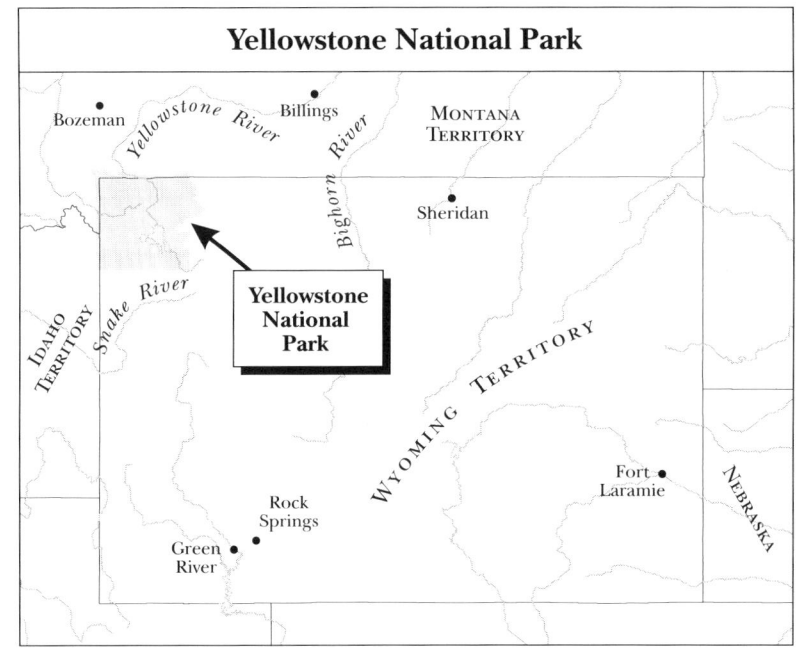

feet) into the sky. Nearby are many natural prismatic springs that are known for their vivid display of colors. Because of both its biodiversity and its unique landscapes, Yellowstone has been declared a World Heritage Site by the United Nations Educational, Scientific, and Cultural Organization (UNESCO).

The numbers of people visiting Yellowstone each year were relatively small until the early 1880's, when the Northern Pacific Railroad built a station in Livingston, Montana, near the northern entrance to the park. Tourism increased steadily after that, and by 1915 roughly one thousand automobiles were entering the park per year. Tourism declined during both World War I and World War II, but efforts to add visitor facilities and modernize the existing ones throughout the 1950's and 1960's were followed by significant increases in visitors. Throughout the 1960's and 1970's yearly attendance at Yellowstone continued to climb, and by the early twenty-first century some two million people were visiting the park every year. An ongoing challenge at Yellowstone, as at all national parks, is that of protecting the park's ecosystems while allowing human access to the park's lands and recreational opportunities.

In the late 1980's tragedy struck the park in the form of uncontrolled forest fires; an estimated 35 percent of the park's total land area was burned. The damage was so great that the entire park was temporarily closed in September, 1988. Prior to the disastrous Yellowstone fires, the National Park Service had taken a suppression approach to the management of natural fires (such as those caused by lightning) in the national parks. Policy makers began to reevaluate this approach, however, as they realized that the suppression of small fires for decades had allowed the growth of large amounts of fuel to sustain larger, devastating fires. In addition, it became clear that prescribed burning in controlled doses is beneficial for the life of a forest. Fire management policies were subsequently changed to reflect this new knowledge.

*Kathryn A. Cochran*

FURTHER READING

Garrott, Robert A., P. J. White, and Fred G. R. Watson, eds. *The Ecology of Large Mammals in Central Yellowstone: Sixteen Years of Field Studies.* Burlington, Mass.: Academic Press, 2009.

Heacox, Kim. *An American Idea: The Making of the National Parks.* Washington, D.C.: National Geographic Society, 2001.

Reinhart, Karen Wildung. *Yellowstone's Rebirth by Fire: Rising from the Ashes of the 1988 Wildfires.* Helena, Mont.: Farcountry Press, 2008.

Sellars, Richard. *Preserving Nature in the National Parks: A History.* New ed. New Haven, Conn.: Yale University Press, 2009.

## Yosemite Valley

CATEGORIES: Places; preservation and wilderness issues
IDENTIFICATION: Scenic valley of the Merced River located on the western slope of the Sierra Nevada in east-central California
SIGNIFICANCE: The Yosemite Valley was part of the first reserve set aside by the United States as an area to be preserved for future generations, a concept that soon thereafter led to the establishment of a national park system.

Caucasians first entered Yosemite Valley in 1851 during conflicts between California gold miners and local Native Americans. In 1864, after lobbying by early conservationists, U.S. president Abraham Lincoln signed the Yosemite Land Grant, which gave Yosemite Valley and an area a few miles to the south called Mariposa Big Tree Grove—15,900 hectares (39,200 acres) of federal land in all—to the state of California as a reserve to be used for public enjoyment and recreation. The state-supervised reserve was the first area specifically set aside by the United States to be preserved for all future generations. Its inception planted the seed for the national park system, although Yellowstone, not Yosemite, was the first site officially designated a national park, in 1872.

Despite the protection provided by the 1864 landgrant legislation, the floor of Yosemite Valley was used for commercial purposes, including plowing and orchard planting, timber cutting, and grazing. The unprotected, high-mountain country surrounding the valley was also logged and grazed. American preservationist John Muir, while exploring the area in the late 1860's, became concerned about these disturbances to the natural landscape. For the next two decades, he publicized his concerns and worked to preserve the high country.

Muir's efforts paid off in 1890 when some 377,000 hectares (932,000 acres) of the high country sur-

*Yosemite Valley, with Half Dome in the distance.* (Courtesy, PDPhoto.org)

rounding Yosemite Valley gained federal protection through the establishment of Yosemite National Park. However, the valley itself and Mariposa Grove remained under California's jurisdiction. Muir and others then worked to get Yosemite Valley transferred from state to federal government jurisdiction in order to protect it and consolidate the public holdings into a single, unified national park. To help rally public support for Yosemite Valley and other land in the Sierra Nevada, Muir and others founded the Sierra Club in 1892.

In 1903 Muir persuaded President Theodore Roosevelt that the valley needed federal protection, and in 1906 California ceded the area back to the federal government. Yosemite Valley and Mariposa Grove thus became part of Yosemite National Park. There was a cost, however: The overall size of the park was reduced to about 308,000 hectares (761,000 acres), and private mining and timber holdings in the park were excluded from restrictions. In 1913, despite the opposition of Muir and other conservationists, the U.S. Congress approved a project to dam and flood Hetch Hetchy Valley in the northwest corner of Yosemite National Park to create a reservoir to supply water to the city of San Francisco.

Since its inception as a state park in 1864, Yosemite has been a magnet for tourists. By the early years of the twenty-first century, the national park was attracting millions of visitors—and their automobiles—each year. Minimizing the environmental damage that can be caused by such high numbers of visitors is an ongoing problem for Yosemite and for other national parks.

*Jane F. Hill*

## Further Reading

Perrottet, Tony. "John Muir's Yosemite: The Father of the Conservation Movement Found His Calling on a Visit to the California Wilderness." *Smithsonian* 39, no. 4 (July, 2008): 48-55.

Radanovich, Leroy. *Yosemite Valley.* Charleston, S.C.: Arcadia, 2004.

## *Zapovednik* system

CATEGORY: Preservation and wilderness issues
IDENTIFICATION: System of nature reserves established in the former Soviet Union
SIGNIFICANCE: The reserves in the *zapovednik* system served two different purposes under the former Soviet Union, with some used as sites of preservation and research and others used mainly for agricultural experimentation. Since the breakup of the Soviet Union in 1991, the reserves have faced difficult times, but some have gained protection and funding after being designated World Heritage Sites or biosphere reserves.

The establishment and administration of nature reserves has generally mirrored the development of biology, ecology, and related sciences. The systematic development of nature reserves with a scientific basis accompanied the developments in biology of the nineteenth century. During that time, Russian biologists related English evolutionist Charles Darwin's evolutionary theories to the environment. From such early scientific studies came a recognition of the importance of natural areas and their preservation.

In the first decade of the twentieth century, Russian zoologists developed the idea of nature reserves, called *zapovedniki*, that would be dedicated to the protection of entire ecosystems. Over the course of the century the popularity of and support for nature reserves in general waxed and waned in the Soviet Union, often influenced as much or more by the political and economic climate as by scientific advances. As was also true in other countries, environmental concerns were seldom of prime importance in the Soviet Union, a giant state that struggled for its own survival for more than seven decades before disintegrating into separate republics in 1991.

In January of 1919, while the new Soviet government was struggling for its existence, agronomist Nikolai N. Pod"iapol'skii proposed the establishment of the regime's first *zapovednik* at Astrakhan. The first five years of the Soviet Union also saw the organization of the All-Russian Society for the Protection of Nature, a volunteer conservation organization that had the effect of enhancing environmental awareness among citizens.

In 1921 Soviet leader Vladimir Lenin signed legislation titled On the Protection of Monuments of Nature, Gardens, and Parks. This empowered the Ministry of Education to declare parcels of nature having special scientific, cultural, or historical value as *zapovedniki*. Between 1919 and 1932 a total of 128 *zapovedniki* were created, with a total area of 12.6 million hectares (31 million acres). The reserves represented 0.56 percent of the total area of the Soviet Union. They varied in size from parcels smaller than 100 hectares (247 acres) to a few of more than 1 million hectares (2.47 million acres) each. Most were in western Russia, Ukraine, or the Caucasus, with a smaller number in Siberia and the far east on the Pacific coast.

The *zapovedniki* were administered by two separate government agencies with different philosophies and sets of goals. The Ministry of Education maintained relatively pristine reserves for their aesthetic properties and as sites for preservation and scientific research. During the 1920's these reserves were utilized by several important Russian ecologists who pioneered studies in such areas as productivity, trophic relationships, and predator-prey interactions in ecosystems. Unlike the national park system in the United States, the *zapovedniki* run by the Ministry of Education did not give great consideration to tourism.

Nature reserves under the management of the Ministry of Agriculture, in contrast, were maintained primarily as centers of agricultural production and experimentation. The Soviets sought to discover scientific management policies that would maximize yields of timber, fur, and other products of value to the economy. These goals compromised conservation efforts.

The emergence of Joseph Stalin as dictator of the Soviet Union in 1929 and the outbreak of World War II in 1939 had disastrous effects on conservation efforts in general and on the *zapovedniki* in particular. Stalin disbanded the Society for the Protection of Nature and introduced a vigorous program of industrialization. Lip service was paid to conservation, but polluting industries were allowed to operate with few regulations.

During this same period, the conservation movement in the Soviet Union was subjected to a fate similar to that of the field of genetics: Scientific principles were abandoned in favor of incorrect, unsupported ideas favored by Marxist theorists. As a result, many unspoiled nature reserves that had been established during the 1920's were dismantled or converted into agricultural enterprises. By 1952 only 40 reserves with a total of 1.5 million hectares (3.7 million acres) were left; this represented just 12 percent of what had existed before the war.

Under Leonid Brezhnev, first secretary of the Communist Party during the 1960's, a new interest in conservation emerged. An improved economy and a reduction in restrictions on individual freedom led to an increase in tourism among Soviet citizens, and appreciation of natural areas grew as international tourists also sought these places out. By 1981 nature reserves had become popular with the Soviet public, and the number had grown to 129, surpassing the number that had existed before World War II.

After the breakup of the Soviet Union in 1991, new challenges emerged. Faced with a legacy of widespread pollution and near bankruptcy, the former states of the Soviet Union struggled to maintain themselves. Many conservationists feared that the environment in general, including the *zapovednik* system, would suffer. The biodiversity of the reserves was threatened both by a lack of funds and by impoverished local people who destroyed the flora and fauna in order to survive. Most observers believed that the assistance of outside agencies would be required to prevent deterioration of the system. By 2010 the number of *zapovedniki* had shrunk to 101. Some of the remaining sites, however, had begun to receive protections and aid with funding through their designation by the United Nations Educational, Scientific, and Cultural Organization (UNESCO) as World Heritage Sites or biosphere reserves.

*Thomas E. Hemmerly*

FURTHER READING

Oldfield, Jonathan D. *Russian Nature: Exploring the Environmental Consequences of Societal Change.* Burlington, Vt.: Ashgate, 2005.

Schwartz, Katrina Z. S. *Nature and National Identity After Communism: Globalizing the Ethnoscape.* Pittsburgh: University of Pittsburgh Press, 2006.

Weiner, Douglas R. *A Little Corner of Freedom: Russian Nature Protection from Stalin to Gorbachev.* Berkeley: University of California Press, 1999.

_____. *Models of Nature: Ecology, Conservation, and Cultural Revolution in Soviet Russia.* 1988. Reprint. Pittsburgh: University of Pittsburgh Press, 2000.

# BIBLIOGRAPHY

Beach, Ben, et al., eds. *The Wilderness Act Handbook.* 5th ed. Washington, D.C.: Wilderness Society, 2004.

Campaign for America's Wilderness. *People Protecting Wilderness for People: Celebrating Forty Years of the Wilderness Act.* Washington, D.C.: Author, 2004.

Carey, Christine, Nigel Dudley, and Sue Stolton. *Squandering Paradise? The Importance and Vulnerability of the World's Protected Areas.* Gland, Switzerland: World Wide Fund for Nature International, 2000.

Chape, Stuart, Mark D. Spalding, and Martin D. Jenkins, eds. *The World's Protected Areas: Status, Values, and Prospects in the Twenty-first Century.* Berkeley: University of California Press, 2008.

Dawson, Chad P., and John C. Hendee. *Wilderness Management: Stewardship and Protection of Resources and Values.* Boulder, Colo.: WILD Foundation, 2009.

Duncan, Dayton, and Ken Burns. *The National Parks: America's Best Idea—An Illustrated History.* New York: Alfred A. Knopf, 2009.

Lockwood, Michael, Graeme Worboys, and Ashish Kothari, eds. *Managing Protected Areas: A Global Guide.* Sterling, Va.: Earthscan, 2006.

Lowry, William R. *Dam Politics: Restoring America's Rivers.* Washington, D.C.: Georgetown University Press, 2003.

Riley, Laura, and William Riley. *Nature's Strongholds: The World's Great Wildlife Reserves.* Princeton, N.J.: Princeton University Press, 2005.

Scarce, Rik. *Eco-Warriors: Understanding the Radical Environmental Movement.* Updated ed. Walnut Creek, Calif.: Left Coast Press, 2006.

Scott, Doug. *The Enduring Wilderness: Protecting Our Natural Heritage Through the Wilderness Act.* Golden, Colo.: Fulcrum, 2004.

Sellars, Richard. *Preserving Nature in the National Parks: A History.* New ed. New Haven, Conn.: Yale University Press, 2009.

# CATEGORY INDEX

ACTIVISM AND ADVOCACY
    Global ReLeaf, 46
    Leopold, Aldo, 67
    Lovejoy, Thomas E., 70
    Mather, Stephen T., 71
    Roosevelt, Theodore, 99
    Sierra Club, 105
    Wilderness Society, 128

AGRICULTURE AND FOOD
    Aquaculture, 4
    Conservation Reserve Program, 19
    Controlled burning, 20
    Soil conservation, 109

ANIMALS AND ENDANGERED SPECIES
    Arctic National Wildlife Refuge, 6
    Bonn Convention on the Conservation of Migratory Species of Wild Animals, 12
    Convention Relative to the Preservation of Fauna and Flora in Their Natural State, 21
    Endangered species and species protection policy, 31
    Fish and Wildlife Service, U.S., 42
    Grand Coulee Dam, 50
    Great Swamp National Wildlife Refuge, 51
    International Union for Conservation of Nature, 61
    Migratory Bird Act, 72
    Wildlife refuges, 131

ECOLOGY AND ECOSYSTEMS
    Antarctic and Southern Ocean Coalition, 2

FORESTS AND PLANTS
    Controlled burning, 20
    Deforestation, 25
    Endangered species and species protection policy, 31
    Global ReLeaf, 46
    Logging and clear-cutting, 68
    National forests, 76
    Sustainable forestry, 116

HUMAN HEALTH AND THE ENVIRONMENT
    Environmental law, international, 37

LAND AND LAND USE
    Alaska National Interest Lands Conservation Act, 1
    Conservation Reserve Program, 19
    Environmental impact assessments and statements, 36
    Greenbelts, 53
    Habitat destruction, 54
    Land-use planning, 65
    Mine reclamation, 73
    Wilderness areas, 126
    Wildlife refuges, 131

ORGANIZATIONS AND AGENCIES
    Antarctic and Southern Ocean Coalition, 2
    Fish and Wildlife Service, U.S., 42
    Global ReLeaf, 46
    International Union for Conservation of Nature, 61
    National Park Service, U.S., 78
    Nature Conservancy, 85
    Sierra Club, 105
    Wilderness Society, 128

PHILOSOPHY AND ETHICS
    Indigenous peoples and nature preservation, 58
    Naess, Arne, 75
    Preservation, 92

PLACES
    Arctic National Wildlife Refuge, 6
    Grand Canyon, 47
    Great Swamp National Wildlife Refuge, 51
    Kings Canyon and Sequoia national parks, 63
    Serengeti National Park, 103
    Yellowstone National Park, 138
    Yosemite Valley, 139

RESOURCES AND RESOURCE MANAGEMENT
    Conservation, 14
    Conservation movement, 16
    Controlled burning, 20
    Environmental law, international, 37
    International Union for Conservation of Nature, 61
    National forests, 76

Soil conservation, 109
Sustainable agriculture, 110
Sustainable development, 114
Sustainable forestry, 116
United Nations Convention to Combat Desertification, 118
Wildfires, 129
World Summit on Sustainable Development, 136

TREATIES, LAWS, AND COURT CASES
Alaska National Interest Lands Conservation Act, 1
Antiquities Act, 3
Bonn Convention on the Conservation of Migratory Species of Wild Animals, 12
Convention Relative to the Preservation of Fauna and Flora in Their Natural State, 21
Environmental law, international, 37
Migratory Bird Act, 72
National Trails System Act, 83
Ramsar Convention on Wetlands of International Importance, 93
Roadless Area Conservation Rule, 98
*Scenic Hudson Preservation Conference v. Federal Power Commission*, 101
*Sierra Club v. Morton*, 107
United Nations Convention to Combat Desertification, 118
Wild and Scenic Rivers Act, 123
Wilderness Act, 125
World Heritage Convention, 134

URBAN ENVIRONMENTS
Land-use planning, 65

WASTE AND WASTE MANAGEMENT
Recycling, 95

# INDEX

Adams, Ansel, 101
Africa
   deforestation, 28
   national parks, 22, 83, 103-105
Agenda 21, 35, 136
Agriculture
   Conservation Reserve Program, 19-20
   slash-and-burn, 21, 26
   soil conservation, 109-110
   sustainable practices, 109-114
Alaska National Interest Lands Conservation Act (1980), 1, 6, 13, 88, 125, 128, 134
Alaska Native Claims Settlement Act (1971), 1
Aldo Leopold National Wilderness, 43
Alligators, 35
Aluminum recycling, 97
American alligator, 35
American Forests, 46
Antarctic and Southern Ocean Coalition, 2-3
Antarctica, preservation, 2-3
Antibiotics, chloramphenicol, 5
Antienvironmentalism, 15
   James Watt, 119
Antiquities Act (1906), 3, 100
Appalachian Trail, 84
Appropriate technology, 55, 115
Aquaculture, 4-6
Arctic National Wildlife Refuge, 6-8
Aswan High Dam, 8-9
Atlantic Treaty System, 2
Audubon Society. *See* National Audubon Society

Bald Eagle Protection Act (1940), 87
Batteries, lead-acid, 97
Beaches, 10-11
Bennett, Hugh Hammond, 109
Biodiversity
   Convention on Biological Diversity, 35
   losses, 28
   wetlands, 121
Biological Dynamics of Forest Fragments Project, 70
Biopiracy, 60
Bioprospecting, 60
Biosphere reserves, 11-12, 91
Biospherical egalitarianism, 75

Bonn Convention on the Conservation of Migratory Species of Wild Animals (1979), 12-13, 35
Boreal forests, 7
Brower, David, 18, 44, 105
Brundtland Commission, 114
Buffalo Protection Act (1894), 87
Buffer lands, 11, 110
Bush, George H. W., 89, 123
Bush, George W., 15, 99, 123

CAFE standards. *See* Corporate average fuel economy standards
Canada, Green Plan, 52-53
Captive breeding, 33
Carrying capacity, 65
Carson, Rachel, 15, 88
Carter, Jimmy, 13-14, 123
Catlin, George, 86
Center for International Environmental Law, 39
Central America, deforestation, 26
China, Three Gorges Dam, 19
Chloramphenicol, 5
CITES. *See* Convention on International Trade in Endangered Species
Civilian Conservation Corps, 15, 18, 87
Clean Air Act (1963), 88
Clean Water Act (1972), 88
Clear-cutting, 68-70, 76, 116
Climate change, international agreements, 39. *See also* Global warming
Clinton, Bill, 15, 98, 123
Colorado River, 105
   dams, 44
Columbia River, 51
Commission for Environmental Cooperation, 39
Comprehensive Environmental Response, Compensation, and Liability Act (1980), 13
Conservation, 14-16
   forests, 116
   movement, 16-19, 100
   versus preservation, 17, 92-93
Conservation easements, 85
Conservation International, 25
Conservation Reserve Program, 19-20
Conservation tillage, 110, 113

Consolidated Edison Company, 102
Contour plowing, 110, 112
Controlled burning, 20-21, 131
Convention on Biological Diversity (1992), 35
Convention on International Trade in Endangered Species (1973), 34, 62, 89
Convention on the Conservation of Migratory Species of Wild Animals. *See* Bonn Convention on the Conservation of Migratory Species of Wild Animals
Convention on Wetlands of International Importance. *See* Ramsar Convention on Wetlands of International Importance
Convention Relative to the Preservation of Fauna and Flora in Their Natural State (1933), 21-22
Coral reefs, 55
Cover crops, 110, 113
Crop rotation, 113
Cross-Florida Barge Canal, 22-24

Dams
   Aswan High, 8-9
   and earthquakes, 9
   Glen Canyon, 44-46, 49
   Grand Coulee, 50-51
   Hetch Hetchy, 56-58
   Tellico, 117-118
DDT. *See* Dichloro-diphenyl-trichloroethane
Debt-for-nature swaps, 24-25
   Thomas E. Lovejoy, 70
Deep ecology, 75
Defenders of Wildlife, 89, 134
Deforestation, 25-29, 54, 116
Department of Energy, U.S., 14
Desertification, 54
   United Nations Convention to Combat Desertification, 118-119
Development gap, 137
Devils Tower National Monument, 3
Dichloro-diphenyl-trichloroethane, 88
   *Silent Spring*, 15
Dinosaur National Monument, 105
Dominion Forest Reserves and Parks Act (1911), 80
Dominion Parks Service, 80
Du Pont Corporation, 129
Duck stamp program, 42, 87, 134
Dust Bowl, 87

Earth Day, 15, 89
Earth First!, 44

Earth Summit (1992), 90, 118, 136
Earthjustice, 99, 106
Earthquakes, 9
Echo Park Dam opposition, 105
Ecological Society of America, 85
Economic growth, 65, 115
Ecosophy T, 75
Ecosystem services, 55
Ecosystems, 14
   riparian, 44, 48-49
   wetlands, 120-123
Ecotourism, 29-31, 91
Electronic waste, recycling, 97
Emerson, Ralph Waldo, 87
Endangered species
   Bonn Convention, 12-13
   IUCN Red List, 62
   policy making, 31-36
   snail darter, 117-118
Endangered Species Act (1973), 34-35, 42, 88, 134
Endangered Species Conservation Act (1969), 88
Endangered Species Preservation Act (1966), 88
Environmental cleanup, mining, 73-75
Environmental Defense Fund, 23
Environmental ethics, Aldo Leopold, 67
Environmental impact assessments, 36-37
Environmental justice, 40
Environmental law, 37-41
   endangered species, 34
   standing, 101-103, 107-109
Environmental policy, 37-41
   endangered species, 31-36
   forests, 98-99
   preservation, 86-90
   and public opinion, 88
Environmental Protection Agency, 18, 88
Environmentalism, 15
Erosion, 20, 109
   beaches, 10
   Nile River, 9
European Diploma of Protected Areas, 41
E-waste. *See* Electronic waste
Extinctions, 31

Farming. *See* Agriculture
Federal Aid in Wildlife Restoration Act (1937), 42
Fires
   management, 20-21, 130, 139
   wild, 129-131, 139

Fish and Wildlife Conservation Act (1980), 13
Fish and Wildlife Service, U.S., 42-43
Fish farming. *See* Aquaculture
Floodplains, Nile Valley, 8
Floods, deforestation, 28
Florida Defenders of the Environment, 23
Food Security Act (1985), 109, 122
Food, Agriculture, Conservation, and Trade Act (1990), 111
Forest Pest Control Act (1947), 88
Forest Reserve Act (1891), 87
Forest Service, U.S., 78
Forests
    deforestation, 25-29
    fire management, 21, 130, 139
    management, 14, 68-70, 76-78, 116-117
    old-growth, 27
    reforestation, 46-47
    Roadless Area Conservation Rule, 98-99

Galápagos Islands, 30
Garbage. *See* Solid waste
Garfield, James, 87
Gasohol. *See* Ethanol
Gene banks, 33
Gila Wilderness Area, 43-44, 67, 126
Glass recycling, 97
Glen Canyon Dam, 44-46, 49, 105
*Global 2000 Report, The* (Council on Environmental Quality), 70
Global Environment Facility, 119
Global ReLeaf, 46-47
Global warming, 40. *See also* Climate change
Gore, Al, 15
Gorillas and ecotourism, 29
Grand Canyon, 44, 47-50, 106
Grand Coulee Dam, 50-51
Grazing, rotational, 113
Great Smoky Mountains National Park, 18, 82, 118
Great Swamp National Wildlife Refuge, 51-52
Green manure, 113
Green Plan, 52-53
Green political parties
    Europe, 90
Greenbelts, 53-54
Greenpeace, 89
Greenwashing and ecotourism, 30
Grinnell, George Bird, 73
Groundwater pollution, 111

Habitat destruction, 54-56
Harkin, John Bernard, 80
Hemenway, Harriet, 73
Hetch Hetchy Dam, 17, 56-58, 92, 105, 140
Hewett, Edgar Lee, 3
Homestead Act (1862), 86
Hunting, 33
    and conservation, 17, 99
    whales, 35

Indigenous peoples
    Alaska, 1
    and nature preservation, 58-60
Industrial Revolution, 86
Insecticides. *See* Pesticides
Integrated pest management, 114
Intergenerational justice, 115
Intergovernmental Panel on Climate Change, 39
International Convention for the Regulation of Whaling (1946), 35
International Environmental Protection Act (1983), 89
International Maritime Organization, 2, 39
International Union for Conservation of Nature, 61-62, 89
International Whaling Commission, 35
Irrigation, 111
Izaak Walton League, 17

Johannesburg Summit. *See* World Summit on Sustainable Development
John Muir Trail, 62-63
Johnson, Lyndon B., 84, 92

Kaibab Plateau deer disaster, 67
Kennedy, John F., 15
Kings Canyon National Park, 63-65

Lacey Act (1900), 34, 42, 87
Lacey, John F., 3
Lake Nasser, 8
Lake Powell, 44
Land clearance, 86
    agriculture, 54
    erosion, 109
Land ethic, 67
Land use
    greenbelts, 53-54
    multiple-use approach, 89
    policy making, 92
    urban planning, 65-67

Law, environmental, 37-41
Leopold, Aldo, 15, 43, 67-68, 128
Little Tennessee River, 117
Logging, 68-70, 116
    deforestation, 26
Lovejoy, Thomas E., 70-71
*Lujan v. National Wildlife Federation* (1990), 108

MacKaye, Benton, 128
*Man and Nature* (Marsh), 14, 86
Man and the Biosphere Programme, 11
Marine Biodiversity Information Network, 3
Marsh, George Perkins, 14, 86
Mather, Stephen T., 17, 71-72
Migratory Bird Act (1913), 72-73
Migratory Bird Hunting and Conservation Stamp Act (1934), 42, 87, 134
Migratory Bird Treaty Act (1918), 42, 73, 89
Mineral King Valley, 107
Minimum Critical Size of Ecosystems Project, 70
Mining, land reclamation, 73-75
*Monkey Wrench Gang, The* (Abbey), 44
Monoculture, 111, 114
Montreal Protocol (1987), 90
Mountain gorillas and ecotourism, 29
Muir, John, 17, 63, 87, 92, 100, 139
    Hetch Hetchy Dam, 57
    Sierra Club, 105
Mulch-till planting systems, 110
Multiple-use land management, 76, 89, 92, 134
Multiple Use-Sustained Yield Act (1960), 76
Municipal solid waste. *See* Solid waste

Naess, Arne, 75
National Acid Precipitation Act (1980), 88
National Audubon Society, 73
National Biological Survey, 71
National Environmental Policy Act (1970), 18, 36, 88
National Forest Management Act (1976), 76, 88, 128
National forests, 76-78
National Marine Sanctuary Program, 35
National monuments, 3, 79, 100
National Park Service, U.S., 17, 71, 78-80
    establishment, 80
    Stephen T. Mather, 71-72
National parks, 76, 80-83
    Grand Canyon, 47
    Kings Canyon, 63-65
    management, 78-80
    Sequoia, 63-65
    Serengeti, 103-105
    Yellowstone, 138-139
    Yosemite, 139-141
National Parks Act (1916), 80
National Trails System Act (1968), 83-85, 128
National Wild and Scenic Rivers System, 123-125
National Wilderness Preservation System, 125-128
National Wildlife Refuge System, 34, 42, 133
National Wildlife Refuge System Improvement Act (1997), 89, 128
Natural capital, 25
Natural resources, policy making, 86-90
Natural Resources Conservation Service, 109
Nature Conservancy, 15, 85-86, 88
Nature reserves, 59, 90-92. *See also* Wilderness areas, Wildlife refuges
    *Zapovednik* system, 141-142
New Deal conservation policy, 87
Nile River, Aswan High Dam, 8
Nixon, Richard, 23, 88
North American Agreement on Environmental Cooperation (1992), 39
North American forests, 26
North American Free Trade Agreement (1992), 39
North American Wildlife Foundation, 52

Obama, Barack, 16
Oceans currents, 2
Ocklawaha River, 24
Omnibus Public Land Management Act (2009), 127, 129
Organic Act (1916), 80
O'Shaughnessy Dam. *See* Hetch Hetchy Dam
*Our Common Future* (Brundtland Commission), 115
Overfishing, 4
Overgrazing, 54

Paper manufacture, recycling, 96
Parks, national, 47, 63-65, 78-83, 103-105, 138-141
Pelican Island, 87, 131
Pesticides, 88, 111
    in aquaculture, 5
Pinchot, Gifford, 14, 17, 57, 76, 87, 100
Pittman-Robertson Act (1937), 42
Plantation forestry, 116
Plastics recycling, 97
Plitvice Lakes National Park, 83
Poaching, 33, 91

Politics
  and environmentalism, 15
  Green parties, 90
Pollution. *See* Water pollution
Polystyrene, 98
Poverty, 55, 137
Power plants, hydroelectric, 44
Predator management, 43, 67
  wolves, 18, 82
Preservation
  Antiquities Act, 3
  versus conservation, 14
  natural resources, 86-90
  wilderness, 92-93, 126-128
Public trust doctrine, 91

Ramsar Convention on Wetlands of International Importance (1971), 91, 93-95, 123
Rangelands, management, 21
Reagan, Ronald, 15, 120
Recycling, 95-98
Red List (International Union for Conservation of Nature), 62
Reforestation, 46-47
Reservoirs, 89
  Lake Nasser, 8
  Lake Powell, 44
Resource recovery, 95-98
Restoration ecology, 116
Right of Way Act (1901), 57
Roadless Area Conservation Rule (2001), 98-99, 126, 129
Rodman Dam, 24
Roosevelt, Franklin D., 15, 18, 22, 87
Roosevelt, Theodore, 3, 14, 17, 73, 87, 99-100, 122, 131, 140
Rowell, Galen, 100-101
Runoff, agricultural, 111

Sagebrush Rebellion, 119
Salinization, soil, 112
Salmon
  aquaculture, 4
  Columbia River, 51
*Sand County Almanac, A* (Leopold), 15, 67
Sarasin, Paul, 61
*Scenic Hudson Preservation Conference v. Federal Power Commission* (1965), 101-103
Scott, Peter, 2, 61
Sea-level changes, 60

Seafood farming. *See* Aquaculture
Seed banks, 33
Sequoia National Park, 63-65, 107
Serengeti National Park, 103-105
Sessions, George, 75
Seville Strategy for Biosphere Reserves, 11
Shrimp farming, 5
Sierra Club, 17, 63, 87, 105-107
  Glen Canyon Dam, 44
  Hetch Hetchy Dam, 57
*Sierra Club v. Morton* (1972), 107-109
*Silent Spring* (Carson), 15, 88
Slash-and-burn agriculture, 21, 26
Snail darter, 35, 117-118
Soil
  conservation, 109-110, 112
  erosion, 20
  fertility, 113
  salinization, 112
Soil Conservation Service, 87, 109
Solid waste recovery, 95-98
South America, deforestation, 28
Soviet Union, *zapovednik* system, 141-142
Species diversity, 131
  Arctic National Wildlife Refuge, 6
Stalin, Joseph, 141
Standing (legal concept), 101-103, 106-109
Stockholm Conference. *See* United Nations Conference on the Human Environment
Stockholm Declaration (1972), 38
Strategic environmental assessments, 37
Strip farming, 110, 112
Styrofoam. *See* Polystyrene
Superfund (1980), 13
Surface Mining Control and Reclamation Act (1977), 74
Sustainable agriculture, 110-114
Sustainable development, 65, 110, 114-116
  World Summit (2002), 136-138
Sustainable forestry, 27, 69, 76, 116-117

Taylor Grazing Act (1934), 87
Tellico Dam, 35, 117-118
Tennessee Valley Authority, 18, 87
  Tellico Dam, 117
Terrace farming, 113
Thoreau, Henry David, 17, 87
Tourism, 29-31
  environmental impacts, 49, 91, 139
Toxic Substances Control Act (1976), 88

Trade Records Analysis of Flora and Fauna in Commerce, 62
Traditional ecological knowledge, 58-60
Trans-Alaska Pipeline, 1
Trash. *See* Solid waste
Truman, Harry S., 87

Udall, Stewart L., 15
UNESCO. *See* United Nations Educational, Scientific, and Cultural Organization
United Nations Commission on Sustainable Development, 136
United Nations Conference on Environment and Development (1992). *See* Earth Summit
United Nations Conference on the Human Environment (1972), 89
United Nations Convention to Combat Desertification (1994), 118-119
United Nations Educational, Scientific, and Cultural Organization, 91
    biosphere reserves, 11-12
    World Heritage Sites, 134-136
United Nations Environment Programme, 39
United Nations population conferences, 90
Urban planning and land use, 65-67

Walt Disney Enterprises, 107
Waste. *See* Electronic waste, Solid waste
Waste management, recycling, 95-98
Water conservation, 112
Water pollution, agricultural runoff, 111
Water Pollution Control Act (1972), 88
Watt, James, 15, 119-120
Weed control, 114
Weeks-McLean Act (1913), 72-73
Wetlands, 120-123
    Ramsar Convention, 91, 93-95
Whaling, 35

Wild and Scenic Rivers Act (1968), 123-125, 128
Wilderness Act (1964), 18, 92, 125-128
Wilderness areas, preservation, 43-44, 67, 92-93, 125-128
Wilderness Society, 128-129
Wildfires, 129-131
Wildlife management, 32-33
    predators, 18, 43, 67, 82
Wildlife refuges, 17, 32, 42, 100, 131-134
    Arctic, 6-8
    Great Swamp, 51-52
Wirth, Conrad L., 81
Wise-use movement, 14
Wolves, 138
    Yellowstone National Park, 18, 82
World Commission on Environment and Development. *See* Brundtland Commission
World Heritage Convention (1972), 134-136
World Heritage Sites, 62, 80, 91
    Grand Canyon, 47
    Serengeti National Park, 104
    Yellowstone National Park, 139
"World Scientists' Warning to Humanity" (Union of Concerned Scientists), 90
World Summit on Sustainable Development (2002), 90, 136-138
World Wide Fund for Nature, 61
World Wildlife Fund. *See* World Wide Fund for Nature

Yard, Robert S., 128
Yellowstone National Park, 16, 80, 129, 138-139
    wolves, 18, 82
Yosemite National Park, 56, 105, 139-141
    Galen Rowell, 101

Zahniser, Howard Clinton, 18
*Zapovednik* system, 141-142